The Population Ecology of Cycles in Small Mammals

The Population Ecology of Cycles in Small Mammals

MATHEMATICAL THEORY AND BIOLOGICAL FACT

JAMES PATRICK FINERTY

Yale University Press
New Haven and London

Published with assistance from the foundation
established in memory of Philip Hamilton McMillan
of the Class of 1894, Yale College.

Set in Monophoto Times Roman type.
Photosetting by Thomson Press (India) Limited, New Delhi.
Printed in the United States of America by
Halliday Lithograph, West Hanover, Mass.

Library of Congress Cataloging in Publication Data

Finerty, James Patrick, 1945-
The population ecology of cycles in small mammals.

Bibliography: p.
Includes indexes.
1. Mammal population. 2. Biological rhythms.
I. Title.
QL708.6.F56 599'.05'24 79-23774
ISBN 0-300-02382-0

10 9 8 7 6 5 4 3 2 1

To

G. EVELYN HUTCHINSON

for suggesting it, encouraging it,
and waiting patiently for it to be written.

Contents

Preface

Oscillatory populations in small mammals have been attributed to several species inhabiting the boreal forest and arctic tundra of the northern hemisphere. Study of these cycles began with excessive enthusiasm in the early part of this century, but an element of skepticism introduced by Cole (1954) caused ecologists to reconsider the subject in the 1950s. At least in some species, the statistical techniques introduced by Moran (1949a) had provided a firm foundation for discussing the reality of cycles. But many ecologists tended to avoid the question of cycles as a subject that could be usefully explored mathematically, while many mathematicians and mathematically oriented ecologists developed interesting new viewpoints that might help to understand this curious phenomenon. Consequently mathematical theory and biological fact became separated, each developing interesting ideas within its own discipline while lacking the supportive verification of its associated discipline.

Logically the first question one must ask is whether cycles actually exist. The evidence presented here seems overwhelmingly affirmative. Once one accepts the strong statistical support for the existence of cycles in some species, he is confronted with the question of why cycles exist in some species in some areas but not in the same or similar species in other areas. Approaching this question necessarily takes one into the fields of biology, mathematics, statistics, climatology, geography, physiology, and biogeochemistry and inevitably suggests one possible vantage point from which to view ecological systems. Certain ecologists (like myself) might fondly call this a Hutchinsonian question. Almost anyone would call it a grand adventure. I hope that the reader will experience a portion of what this adventure has meant to me.

The delightful drawings of cyclic species which appear throughout this volume are from J. J. Audubon's (1854) *Quadrupeds of North America.*

Acknowledgments

The manuscript has benefited from the suggestions of a number of people: Frank A. Pitelka conveyed his broad field experience in Alaska and Scandinavia; Jerry Wolff, Charles Krebs, Mark Williamson, W. A. Fuller, Richard Levins, Lloyd B. Keith, Mark Boyce, John Fox, and Egbert Leigh offered helpful information; Ulla Kasten orally translated Collett's important Norwegian texts; the libraries of Yale University provided consistently superior service to match their consistently superior collections; Mary and Don Poulson offered companionship and shelter in the initial stages of writing; and Eric Charnov and Richard S. Miller made several helpful suggestions for the final manuscript.

The book actually began on the third floor of the Osborne Memorial Laboratories of Yale University in the autumn of 1967 when, while helping me select a topic for a term paper in his famous course on ecological principles, Evelyn Hutchinson discovered that I was experienced in mathematics and was studying computer sciences. He was convinced that computerized statistics, following the lead of Moran (1949 et seq.), would point the way toward an answer to a question that had been becoming increasingly bothersome during the last 25 years. I will not soon forget his excitement when, during the last 15 minutes of the last day of class, I presented the results.

Several years later, while I was searching for a final thesis topic, Evelyn suggested this question for a thesis.

The next major step came in early 1976, while discussing several proposed publications over tea at the Yale Elizabethan Club, the same historic setting in which Evelyn and Lars Onsager worked out the mathematical consequences of the formulation of the logistic equation with a time lag (Wangersky, 1978): Evelyn wondered if there might be enough in this question for a book. My naive assent resulted, some 20 minutes later, in the appearance of the inimitable Jane Isay, editor extraordinaire of the Yale University Press, whose constant encouragement thus began at the inception. Several months later a postdoctoral appointment with the Biology Department of the University of Utah, offered at the request of Maxine Watson, Eric Charnov, and Graham Pyke, provided the indispensable financial support for the project.

Among the people who have the good fortune to experience the influence of great teachers during their formative years are certainly the students of Evelyn Hutchinson. His work and the spirit of his students stand as a testimony to his ceaseless efforts to try to make comprehensible the strange and fascinating universe in which we reside. To him I therefore gratefully dedicate this book.

Systematic and Common Names of Mammals

CARNIVORA

MUSTELIDAE

Martes pennanti	fisher
Martes americana	marten
Mustela vison	mink
Mustela erminea	ermine
Mephitis mephitis	skunk
Gulo luscus	wolverine
Lutra canadensis	otter
Enhydra lutris	sea otter

FELIDAE

Lynx canadensis	lynx
Lynx rufus	bobcat

CANIDAE

Canis latrans	coyote
Canis lupus	wolf
Alopex lagopus	arctic fox
Vulpes spp.	red (colored) fox

PROCYONIDAE

Procyon lotor	raccoon

PINNIPEDIA

PHOCIDAE

Phoca vitulina	hair seal

OTARIIDAE

Callorhinus ursinus	fur seal

RODENTIA

MURIDAE

Lemmus spp.	brown lemming

Dicrostonyx spp.	collared lemming
Ondatra zibethicus	muskrat

LAGOMORPHA

LEPORIDAE

Lepus americanus	snowshoe rabbit

1 *Introduction*

The present aim of the part of ecology with which this book is concerned is therefore largely to uncover possibilities, by any kind of theoretical analysis that proves helpful, and then to see how many of these possibilities are indeed realized in nature.

G. Evelyn Hutchinson (1978)

The observation of fluctuations in biological populations simply reflects the fact that population density is hardly ever exactly its average value, but is sometimes larger, sometimes smaller. If these fluctuations are to be called *oscillations*, however, certain restrictions must be met, namely that if at some time t, N, the size of the population, is greater than the mean value of the population over time, there will be a propensity at some future time for N to be less than its mean value (Moran, 1954). This is not simply to say (as Cole, 1954, suggests) that the population does not remain at a constant size, but rather that the population fluctuates around a constant size. If the term *cycle* is to be applied, more stringent criteria will be necessary (Keith, 1963): (1) a succession of events (an increase above the mean followed by a decrease below the mean, and finally a return to the mean) must be completed; (2) the succession of events must be recurrent; and (3) the recurrence must be at relatively regular time intervals. In addition, since science is confined to the study of things that can be measured, the amplitude of the cycles must be large enough to distinguish them from random fluctuations.

Several examples of oscillatory populations exist and some of these have been termed cyclic. One immediately thinks of 17-year cicadas (Alexander and Moore, 1962) and larch budmoths (Baltensweiler, 1971; Baltensweiler et al., 1977; Auer, 1977). But the most extraordinary examples are in small mammals. These small mammal cycles can be separated into two categories which can be considered characteristic of the northern hemisphere: the 10-year cycle of the North American boreal forest, notably snowshoe hares, muskrats, and their predators; and the 4-year cycle of the arctic tundra in Eurasia and North America, notably lemmings, and perhaps other microtines, and their predators. The existence of these "cycles" has raised many challenging questions over the years: Why do many of the populations settle about a constant mean instead of dying out or increasing infinitely (Moran, 1954)? Why do the cycles in numbers of lynx occur across Canada in almost perfect synchrony (Moran, 1953a)? Why does the 10-year cycle appear to be restricted to the northern coniferous forest of North America and its ecotonal communities (Keith, 1963)? How is it that the same sequence of cyclic events is often superimposed over differences among species, habitats, and geographical areas (Tamarin, 1978). Does the existence of cycles of similar periods among such diverse organisms reflect something inherent in complex biological systems, or perhaps a worldwide control factor (MacLulich, 1937)?

Before any of these questions can be seriously entertained, a more basic question must be answered: Are these oscillations periodic in a mathematical sense, or can they be considered a series of random variations? The importance of answering this question definitively cannot be overemphasized: twentieth-century ecological studies have been distinguished by a lengthy history of often questionable attempts to connect many phenomena by correlation analysis under the assumption that within the physical environment, or the biological system, are factors that tend to induce the observed population cycles. A tendency to accept "periodicities" without a critical view toward what constitutes a cycle developed, and a good deal of research was undertaken to discover causal factors for these cycles before they were shown to exist (Butler, 1953).

In 1951, Cole, following the lead of Palmgren (1949), suggested that these cycles could possibly be explained as random fluctuations, and he stressed the necessity of eliminating this possibility before attempting "more complicated and indirect explanatory hypotheses" (Cole, 1954). Cole postulated that all population cycles might be most easily explained as random series. His argument was simple: Given N random numbers, all different (so that no ties can occur), all independent, and all with an equal probability of occurring, a peak can be defined as any observation that is larger than the one preceding it, and larger than the one following it. Therefore, given Z_1, Z_2, and Z_3, a peak occurs when

$$Z_1 < Z_2 > Z_3$$

Given these three numbers, six different permutations are possible, and two of these will result in a peak. Thus the probability of observing a peak is two in six, or one in three. For N numbers, there are $(N - 2)$ groups of three, so the average number of peaks will be $(N - 2)/3$. Since a cycle can be defined as the distance between two peaks, there will be

$$\frac{N-2}{3} - 1 = \frac{N-5}{3}$$

cycles. The mean length of a cycle will be the length of the series divided by the mean number of cycles, or

$$\frac{N}{(N-5)/3} = \frac{3N}{N-5}$$

Defining a *dominant peak* as one that is higher than its neighbors, the argument can be repeated. The probability of discerning a dominant peak will be 1/3; there will be $(N-2)/3 - 2 = (N-8)/3$ groups of these peaks, giving

$$\frac{1}{3}\frac{N-8}{3} = \frac{N-8}{9}$$

dominant peaks; there will therefore be

$$\frac{N-8}{9} - 1 = \frac{N-17}{9}$$

longer cycles, each of mean length

$$\frac{N}{(N-17)/9} = \frac{9N}{N-17}$$

As N becomes larger, the short cycle approaches a mean length of 3, and the longer cycle a mean length of 9. This led Cole to suggest that the cycle length could be increased by choosing only prominent or conspicuous peaks, and he concluded that the evidence for real cyclicity in animal populations was highly questionable.

Although Cole's analysis injected some long-overdue skepticism into the subject of cyclic phenomena, many criticisms can be legitimately leveled against his method. His definition of a peak seems somewhat restrictive by not allowing for situations where values are fluctuating randomly around a trend; and it neglects amplitude, so that obviously inconsequential "ripples" will be counted. Second, Cole does not admit the distinction between a *mean cycle length* of 3 to 4 years and a *regular period* of 3 to 4 years, the difference being the magnitude of the variance of cycle length. The concept of mean distance between peaks is not particularly useful, because it is possible to construct a nonoscillatory equation with mean distance arbitrarily large, for example a moving average of random numbers with positive weights (Moran, 1954). There is also the question of synchrony: If peaks occur fortuitously, why would various regions of a country show peaks in the same year (Butler, 1953), and why would two species that are unrelated to one another in size, life span, or reproduction characteristics fluctuate at approximately the same frequency despite the fact that the time unit for all the data considered is 1 year (Siivonen, 1954b)? It is difficult to accept random effects, either local or intrinsic to the biological systems, as an explanation for the widespread synchrony.

Despite all these problems, most biologists were led to believe that all short-term oscillations are random, and that longer ones may also be random (see, e.g., *Cold Spring Harbor Symposium in Quantitative Biology*, Vol. 22, 1957). The task seemed to be to find a way of making more objective statements concerning periodicity which would not be plagued with these many difficulties. Such a method had, in fact, been developed and used for some time by many investigators, and was introduced into ecology by Moran (1949a). This method involves the *autocorrelation function.* This function is a measure of the goodness of fit, or degree of correlation, between a series as given and the same series when initiated after a given time lag. The value of the autocorrelation coefficient at time lag zero is $+1$, signifying the obvious fact that a series is an exact duplicate of itself. For a white noise process, the expected value of the autocorrelation coefficient is 0 for time lags greater than zero. If a series has a regular periodicity such as that of a sine curve, displacing the curve by multiples of one half of a period produces the opposite of the original curve: a maximum in one curve is matched by a minimum in the other, and the autocorrelation coefficient is -1; displacing the original curve by unit multiples of the period exactly duplicates the original curve, yielding a correlation of $+1$ (see fig. 1). Values between $+1$ and -1 define the degree of correlation, either positive or negative. For any periodic series a large positive correlation will be observed when the maxima of the series correspond exactly, and a large negative correlation will result when the maxima of one series correspond with the minima of the other. For a regular periodic phenomenon the function will oscillate at regular intervals, the interval being determined by the frequency of oscillation of the original data. If the observations are oscillatory in the manner defined above, whether regularly periodic or not, there will be at least one negative autocorrelation coefficient of significance of some order (Moran, 1954), but no necessarily regular oscillation of the autocorrelation function.

The advantages of the autocorrelation function are obvious. The problem of finding an objective definition of a peak is eliminated since the series is considered as a whole rather than as individual fluctuations. The

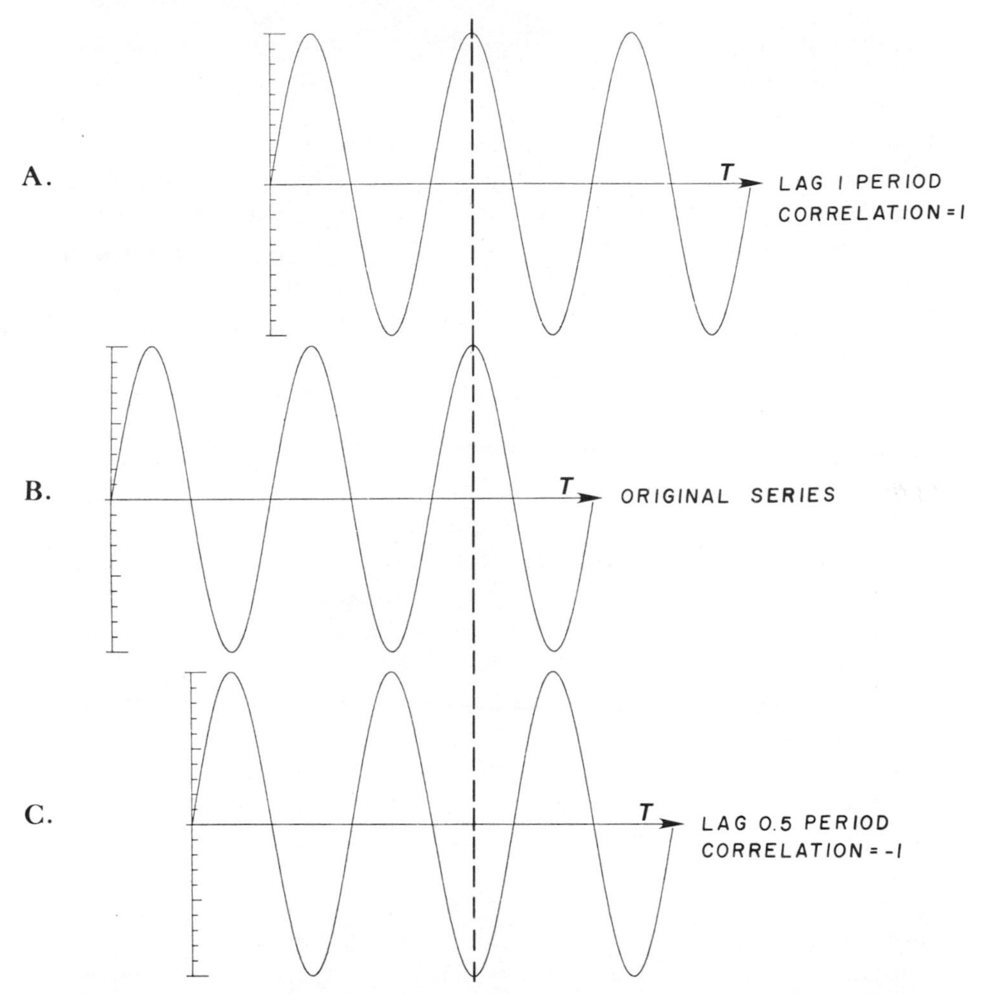

Figure 1. Visual representation of autocorrelation. As a series is displaced by single time units and then compared with the original series, the comparison may be exact, as between A and B, yielding a correlation of + 1, or the comparison may be exactly reversed, as between B and C, yielding a correlation of − 1. The degree of correlation between the two series for other time lags will lie between these extremes. If the original series is truly oscillatory, the autocorrelation function will peak with the same period as the original series.

autocorrelation function effectively redefines "regular peak" in an exact way: a regular peak will result in a significant correlation for the whole series at a given time lag. In addition, the problem of average versus exact cycle length is minimized: the larger the variance between relatively regular peaks, the smaller will be the autocorrelation coefficient.

It is possible to make even more powerful statements concerning series' periodicity by employing *spectral analysis*, a direct extrapolation from correlation analysis. As Cole (1954) observed, any time series of finite length can be expressed with reasonable fidelity as a sum of sine and cosine terms; this is Fourier analysis. Cole interpreted this fact as a barrier to identification of a systematic oscillation, but if a real periodicity exists, one sine or cosine term will dominate the series, and the Fourier coefficient of that term will be significantly larger than the coefficients for the other trigonometric terms. A spectrum essentially converts a series into component sine and cosine terms at various frequencies and measures the contribution of each frequency to the series. For a white noise process this spectrum will be a constant. But if one frequency, or region of frequencies, dominates the series (i.e., if a real periodicity exists), a significant peak will appear in the spectrum; in spectral terminology, most of the power of the spectrum will be in the region of that frequency. An oscillatory autocorrelation function may be due to large covariances between neighboring data values rather than to truly oscillatory data. The spectrum helps to distinguish these two possibilities, because an oscillating autocorrelation function will only produce a spectral peak at the frequency of oscillation if the amplitude of the component sine wave is significantly large (Jenkins and Watts, 1968).

By combining the complementary advantages of correlation analysis and spectral analysis, it will be possible more objectively to ascertain the existence of cyclic population fluctuations. Once a list (not necessarily definitive) of unequivocally oscillatory species is established, it will be possible to reexamine proposed causal connections between various populations, as well as to suggest models for the fluctuations, in an attempt to determine what aspects of the mass of theory concerning these curious cycles need to be seriously entertained. When the theory becomes especially complicated, the reader may be encouraged to continue, as I myself often have been, by remembering that the widespread occurrence of cycles and the regularity of their periods suggest that, from all this complexity, simplicity may emerge. In the process of examining various theories it may occasionally become apparent that there is a great deal of theory with dubious applicability to population cycles, or to real situations in general—yet each theory has had a role in helping to uncover causal mechanisms for these phenomena and thus deserves consideration. After sifting through these theories it will be time, following Professor Hutchinson's suggestion, to try to assess "how many of these possibilities are indeed realized in nature." No "answer," as such, is attempted. But if, at the end, a clearer picture of what we know emerges, and with this picture some idea as to the *nature* of the answer, the purpose of this book will have been realized.

2 *The Perspective of History*

In 1924, Charles Elton, in a paper noteworthy for its scope, called attention to a situation that many had recognized but no one had considered in a rigorous, scientific way: the existence of regular long-term periodic fluctuations in natural animal populations. What Elton documented in this paper, and what he continued to document for decades afterward through his own work and through the creative efforts of the Bureau of Animal Population which he directed at Oxford, opened an area of speculation which continues to inspire and puzzle natural scientists. Unfortunately, the excitement of the suggestion that some mammals and birds fluctuate periodically over arctic and boreal regions in the northern hemisphere, combined with the difficulties of obtaining long-term records to verify proposed theories, have precipitated as much confusion as insight into mechanisms generating these cycles, and it now seems appropriate to try to evaluate what we know, and what we do not know, about regular fluctuations in natural populations.

The history of the subject makes it necessary to begin by defining what we mean by "cycles" in natural populations. The background of the use of this term by biologists is reviewed by Keith (1963). By a cycle with a period of, say, 4 years we shall mean that the highs (zeniths) and lows (nadirs) of the fluctuation are temporally spaced in such a way that the time distance between zeniths (or nadirs) clusters around 4 years to a degree that would not be expected by chance; the distance may be 3 years on occasion, or sometimes 5 years, but is 4 years more often than not. Nothing is suggested about the amplitude of the peaks or nadirs. Later we shall see how Moran (1949a) gave this definition a rigorous statistical form which makes the existence of the cycle, although not necessarily its non-existence, unequivocal.

Long-term mammalian population cycles have been suggested for several regions, but two cycles stand out clearly: a 4-year cycle, associated with the barren arctic tundra, and a 10-year cycle, associated with the northern boreal forest. The clearest and earliest examples of 4-year cycles are *Lemmus lemmus* (lemming), *Vulpes vulpes* (red fox), and *Alopex lagopus* (arctic fox) in Norway. For 10-year cycles *Lepus americanus* (snowshoe hare) and *Lynx canadensis* (Canadian lynx) are outstanding. Many other examples of oscillatory populations will be cited later.

Documentation for the nineteenth century (Elton, 1924; MacLulich, 1937; Elton and Nicholson, 1942a and 1942b; Elton, 1942) clearly reveals both 4- and 10-year cycles. Earlier documentation is less clear. If cycles have existed for many centuries, or much longer (as suggested by Elton, 1924), when were they first noticed? What brought them to notice? If the regularity of these population fluctuations was not recognized, why not?

Because of their conspicuous mass migrations, Scandinavian lemmings have received a good deal of attention in the literature. The first generally accepted reference to lemmings is an old manuscript, attributed to the latter half of the thirteenth century, in which the Biblical locust plague in Egypt is compared by a Norwegian translator to a lemming plague (Collett, 1895). Lemmings do not appear to have been mentioned before this (with the possible exception of an Icelandic saga; Marsden, 1964). This is probably so because the lemming habitat in Norway is typically high above arable and grazing land and is generally removed from settlements, and the only time their presence would be obvious would be during "lemming years," when they were thought of as pests—unclean rodents to be exterminated—and certainly not worthy of celebration in literature (Marsden, 1964).

After the mention of lemmings in the thirteenth century there seems to be little interest in these rodents until the sixteenth century. Then, from the Lapland area, came the story of many-colored mice that fall from the sky, a story conveyed to Europe as a historical coincidence. During the Reformation the Norwegian province of Trondelag was a stronghold of Roman Catholicism, and from here Archbishop Erik Walkendorfer, and later Archbishop Olav Englebrektsson, went to Rome. Both of these archbishops were acquaintances of a Bavarian geographer and mathematician named Jakob Ziegler, who, in a book published in 1532, related the archbishops' stories describing lemmings falling from the clouds and either causing pestilence or being eaten by ermines (Collett, 1911–12; Marsden, 1964). But Olaus Magnus, Archbishop of Uppsala, writing in 1555a and 1555b, was the first to ascribe a number to these migrations. In his section concerning ermines, *Mustela erminea* ("Lekat Gothice vel hermelini"), he showed an illustration of lemmings falling from the clouds along with a scene from his map of 1539 depicting two ermines (one in a trap) each with a captured lemming (fig. 2), and he stated: "Hae autem bestiolae [i.e., ermines] singulo *quoque triennio*, ut in plurimum nimia comestione ad maximum lucrum mercatorum ampliantur, ac prolongantur in pelle . . . ," which translates as "But these small beasts, for the most part, every three years for the merchants' exceeding great gain, grow to have their fur very long, because they eat so much. . . ." He continued,

Figure 2. First known depiction of lemmings from Olaus Magnus, 1555b.

paraphrasing Ziegler nearly verbatim, that this increase in ermine size is due to the fact that lemmings fall from the clouds and are eaten by ermines. Thus at least as far back as the mid-sixteenth century there was a sense of regular fluctuations concerning furbearers and their prey.

It is of some historical interest to note that the first record of the regular flux of a prey species is connected to cycles in a predator species that is valued for its fur. We might speculate that the Scandinavian counterpart of Britain's Hudson's Bay Company, that revealing source of information for Canadian fur cycles (to be discussed), provided the first concrete evidence of periodicity in Scandinavian populations, and that it took an archbishop-naturalist, as aware of the *maximum lucrum mercatorum* as any trader, to first record this! It is also interesting to note that the idea of short-term cycles in small mammals appeared in other areas of northern Eurasia. In 1760, Johann Gmelin, describing earlier conversations with intelligent native hunters in Siberia, observed the regularity of the disappearance and reappearance of lemmings and foxes: "Quam regionem deserverunt, eam post tres quatorque annos repetunt. . . ." ". . . they repeat each three and four years." Here again the lemming pest is connected to a predatory furbearer (see also Braestrup, 1941).

Oddly, this sense of regularity does not appear to be mentioned by most other later writers. Erich Pontoppidan, Bishop of Bergen, writing in 1751 (Eng. trans. 1755), and referencing Olaus Magnus, said that the "plague" of lemmings occurs "about once or twice in twenty years." When the plague comes "it only rages here and there in certain districts at a time." Additionally, he noted that "in Sogne Fiordens Fogderie, in this diocese, it happens every third or fourth year, that a few lemmings are seen here, yet but a few, and cannot do much harm." This corresponds to the observation that mass migrations are not a consistent feature of lemming abundances and do not occur at every cycle peak in every area (Elton, 1942; Collett, 1898, in Wildhagen, 1952). Again the lemming flux is connected to predators, since Pontoppidan connected the increased lemming populations with good hunting of bear, fox, marten, and other large animals that devour lemmings. The mass migrations seem to represent the outcome of a strong population peak with large-scale dispersal, whereas the actual cycle, on a local scale, is probably short-term. This masking of the underlying cycle by the more conspicuous migrations might explain why Pontoppidan did not consider lemmings to be periodic, even though the concept of natural cycles was probably circulating in Scandinavian intellectual groups at the time. Pehr Kalm, a Danish naturalist who (somewhat against his will) was sent by Linnaeus to collect North American plant specimens in October 1747, was in Philadelphia during an outbreak of 17-year cicadas in 1749. By examining a parish register of a diligent Swedish rector living near Philadelphia, he determined that for the years 1702–19, 1715 was the only year of cicada outbreak in the area; and by examining, at the suggestion of Benjamin Franklin, the Philadelphia records of an Englishman named Breintnal for 1731–45, he determined that 1732 was the next outbreak; he thus verified the length of the cycle, a considerable accomplishment in view of what we now know to be the complicated nature of cicada broods. This discovery was published in 1756, and was used by Linnaeus (1758) to describe *Cicada septendecim*. The fact that Linnaeus discussed cicada cycles but did not mention regular fluctuations in lemmings suggests that the lemming's cyclic nature was obscured by the more conspicuous mass migrations.

In the nineteenth century the uncertainty about periodicity continued (Bowden, 1869). But the concept of periodicity must have existed to some extent because Clark Kennedy, a British traveler on a 6-week trip to the Norwegian Arctic in 1878, noted that the lemmings "appear in vast quantities every three or four years." Collett, who studied the lemming fluctuations in Norway in detail (1895) and provided (1911–12) the data that Elton (1924) used to reveal the cycles, initially made no note of the cycle, observing only that "no rule can be laid down concerning the frequency of prolific and migratory years. The *greatest* [his italics] migrations, which extend down to the most distant lowlands, take place but seldom on the whole . . ." (Collett, 1895). Later Collett (1898; cited in Wildhagen, 1952) pointed out that a mass increase in lemmings occurs on the average every third or fourth year. Again, the emphasis on mass migrations clouded the issue. Elton (1942) independently discovered that there are not migrations at every cycle peak in every area, but there is a migration somewhere in Norway every 3 to 4 years.

In Norway, foxes were considered vermin and were not even hunted for sport (Bowden, 1869), although a sizable fur export trade existed. Like other lemming predators, foxes were assumed to follow lemmings on their migrations (Bowden, 1869), and since the lemmings were not recognized as being cyclic, it is not surprising that cycles in Norwegian foxes were not noted. However, Collett (1895) stressed that the causes that produced a lemming increase "simultaneously affect many other species," both predators and other microtines. The data that demonstrate the fox cycle (Johnsen, 1929) are from foxes brought in for bounty in Norway, reemphasizing the view of foxes as vermin rather than as valuable furs.

The history of long-term ("10-year") periodic fluctuations in Canada is difficult to trace, since written records preceding the arrival of European fur-trading organizations are unknown. Diereville, writing in 1699, recorded that "The Island of St. John (Prince Edward Island) is stated to be visited every seven years by swarms of locusts or field mice alternately—never together. After they ravage the land, they precipitate themselves into the sea." An imaginative account, no doubt, but one that contains the concept of regularity—a concept that we might infer to be derived from native residents of these areas, because observations over a long period would be necessary to estimate the cycle length. A certain Reverend Patterson (1886), residing in Nova Scotia and trying to gather information concerning these plagues for the consideration of skilled naturalists, knowing the work of Diereville, said that "there is no evidence of any such regularity in the visitation of mice, but later writers speak of it as occurring on Prince Edward Island at longer or shorter intervals, and on the mainland it has not been unknown." Therefore, once again there seems to be evidence of a resistance in the nineteenth century to the concept of regularity in mammalian population fluctuations, even though the concept appears earlier. An investigation of the Hudson's Bay Company's London Archives by Elton and Nicholson (1942b) uncovered a statement by Matthew Cocking of Cumberland House on the Saskatchewan River, for July 1776: "Four or five years ago cats [lynx] were very plentiful here . . . now the natives say there are scarce any; this is attributed to the scarcity of rabbits, these being the cats' chief food." Here we have the concept of population extremes but not of cycles. Examining the manuscript journals of Alexander Henry the younger (see Coues, 1897), the fur statistics for the Red River district for 1800–08 show one complete cycle with a conspicuous peak in several furbearers for 1805–06, yet Henry does not mention a cycle. Even David Thompson, astronomer, geographer, explorer, discoverer, and fur trader, who was interested in mathematics and was associated with the scenes of Henry's journals from 1789 to 1812, did not record notice of the 10-year cycle.

The first recorded suggestion of long-term fluctuations (recorded in Elton and Nicholson, 1942b) was made by Peter Fidler in his "Report of the Manetoba District" to the Hudson's Bay Company in 1820: "The cats (lynx) are only plentiful at certain periods of about 8 to 10 years, and seldom remain in these southern parts in any number for more than two or three years." This rather perspicacious observer also noted: "What is rather remarkable, the rabbits are the most numerous when the cats appear"; and in his 1821 report he added that "the martins this winter have been very scarce, but it is generally observed that when this happens the cats become plentiful." Fidler accurately estimated the lynx cycle length, the predator–prey relationship with the snowshoe hare, and the phase relationship to the marten and suggested the possibility of a mutual interaction in the predator community. As Elton and Nicholson (1942b) pointed out: "The statement implies a knowledge of the cycle among resident traders, that must have been the result of observation over more than one cycle." J. Richardson, who made two overland expeditions to the Polar Sea as a surgeon and naturalist in 1819–21 and 1825–27 and subsequently wrote a monumental treatise on North American fauna (Richardson, 1829), did not mention regularity, only noting that lynxes are numerous when there are plenty of hares, that hare populations are decimated by epidemics, and that these result in a "lapse of several years, during which the lynxes were likewise scarce." This is especially interesting since Richardson's expeditions began at York Factory on Hudson Bay in Manitoba and crossed Saskatchewan into Alberta at the time when Fidler was writing. One can only surmise that Richardson was either not alert to the possibility that cycles might exist, or he may not have felt that regular oscillations were well-enough documented to allow assertions of regularity. Even Poland (1892), who published many of the figures from the Hudson's Bay Company which were later used to demonstrate cyclic regularity, made no explicit recognition of this regularity.

Ross (1861), working in the Mackenzie River district for 13 years, recorded the following: "It [lynx] is subject, like most of the other Fur Animals, to periodical migrations, which appear to occur with great regularity in periods of ten years, and which in its case depend on the Hare its principle food"; and concerning martens (*Martes martes*): "The periodical disappearance of this species is quite remarkable. It occurs in decades, or thereabouts, with wonderful regularity, and it is quite unknown what becomes of them. They are not found dead. The failure extends throughout the Hudson's Bay Territory at the same time. And there is no tract, or region to which they can migrate where we have not posts, or into which our hunters have not penetrated." The idea of long-term cycles is by this time firmly established.

Even when cycles were recognized, there seems to have been some variation of opinion concerning cycle length. Russell (1898), who made several expeditions in Manitoba between 1892 and 1894, noted, concerning *Lepus americanus*: "The natives at different posts gave different dates as the year of greatest abundance, so that I am inclined to think that they do not increase and decrease simultaneously over the whole region, but that it is by periods of seven years all seemed agreed." Lynx are said by Russell to have the same period as hares. Preble (1908), writing about the Athabasca–Mackenzie region also gave 7 years as the rabbit cycle period and said the fluctuation is neither regular, nor in all places at all times. Mair and MacFarlane (1908), writing in the same year about an 1899 expedition through the Mackenzie River basin, referred to the lynx as "one of the principal periodic fur-bearing animals which regularly increase or decrease in numbers about every decade." They also asserted that the marten is "probably the most constant of the 'periodic' fur-bearing animals, whose presence in considerable numbers is very largely dependent upon a greater abundance of hare or rabbits, though mice also form an important item of marten diet." The concept extended beyond hare dependence when they discussed mink (*Lutreola vison lacustris* Preble): "although it is but very slightly dependent on the American hare for food, yet it somehow seems to periodically augment and decrease in numbers much in the same way, not perhaps as precise, but still in a remarkably interesting manner. "Not perhaps as precise" seems restrained, indeed, since evidence shows the mink cycle to be quite regular. Skunks, muskrats, and red foxes were also included in the periodic roster. Ernest Thompson Seton (1909, 1911), whose young people's books stimulated the imaginations of many a naturalist-to-be (G. E. Hutchinson, personal communication), was one of the first to publish part of the now-famous data on fur collections of the Hudson's Bay Company, and to note the 10-year periodicity in many species (Poland, in 1892, published much of the data, but apparently without comment). Hewitt (1921), looking at the same Hudson's Bay data, observed that since the prevailing popular concept was for a 7-year cycle for hare and its predators, the observed 9.7-year cycle in fisher must mean that the periodic flux in this furbearer must be independent of diet—probably a somewhat heretical thought for his times. These differences in opinion concerning cycle lengths may be important in explaining causes of cycles; alternatively it has been suggested that the 7-year cycle might represent the distance between one high or low and the beginning of the next in a 10-year cycle (Leopold, 1931).

The shorter 4-year cycle in Canada, comparable to that in Scandinavia, does not seem to be as thoroughly documented as the 10-year flux. This may be because the Labrador and Ungava peninsulas, whose fur data give the clearest evidence of regularity, were not considered to be as productive in furs as other areas and were only exploited later. The Moravian Missionaries, the first to establish trading posts of extended duration in the area, may also have been more interested in the Indians' welfare than in the fur returns. In the eighteenth century Pennant (1784) had noted the short-term cycle in the pine marten (*Martes americana*): "Once in two or three years they come out in great multitudes, as if their retreats were overstocked." Pennant also noted fluctuations and migrations in arctic fox, recording a report from a Mr. Graham that they were only migratory in the Hudson Bay area once in 4 or 5 years, and that they "will at times desert their native countries for three or four years, probably as long as they can find any prey." Hind (1863), writing in Toronto, also observed a recurrent population flux: "It is on this river (St. Augustine River, Labrador) that the curious migration of animals every third or fourth year is particularly observed. The year 1857 was one of these migratory years and during the winter the hunters on the lower part of the St. Augustine, 50 miles from the sea, reaped a rich harvest of otters, martens, foxes." The form of this statement suggests that this regularity was well known. Hewitt (1921) determined that in the arctic fox "numbers appear to fluctuate very considerably over shorter periods than is the case with the more southerly red fox and its color phases." Hewitt calculated a 4.2-year cycle, but said that 4 years is by far the most frequent period.

After Hewitt's (1921) discussion of cycles, a new era in the study of cycles began to emerge as Elton (1924 et seq.) began to expand the valid data base on oscillatory fluctuations and, with his coworkers, to delve into the massive records of the Hudson's Bay Company Archives, and Volterra (1926) began to apply deterministic mathematics to population variation. The output from this era, which continues to this day, has been both exciting and confusing, and it is the output of this era that is the major concern of this study.

A Note on Counting

The question of the degree to which the American Indians and Eskimos recognized regular fur cycles has hardly been studied. The fur traders likely learned something of these cycles from the Indians and Eskimos, but since in some areas it was considered a crime to talk to Indians (Robson, 1752), there may not have been much information transferral at this level. Bison, caribou, and reindeer (Thompson, 1812; Elton, 1942; Bell, 1884) were main sources of food and clothing, and evidence suggests the natives had a sophisticated understanding of their movements, as Thompson (1812) noted when discussing bison movements with Piegan Indians in the winter of 1787. J. K. MacDonald of Winnipeg, after 35 years as a trader with the Hudson's Bay Company, made similar observation when he recorded that the Indians' knowledge of marten movements was so definite that they would go in the winter to head off the wanderers, and would inevitably make contact with them (Seton, 1909). It seems reasonable, therefore, to assume that the Indians would be aware of extreme regular variations in populations of furbearers. The Indians in the early twentieth century took note of the number of embryos in captured female hares during the breeding season as an indicator of abundance for the coming winter (Preble, 1908) and they were aware of oscillations in numbers of births in other cyclic furbearers (Mair and MacFarlane, 1908), but how far into the past this knowledge extended is unknown. The fact that the periodic failure of hares was anticipated by the Indians with foreboding, yet did not inspire them to provide for the future (Preble, 1908), suggests that the Indians did not feel they could accurately predict the crash. Von Wrangell's (1844) picture, in Siberia, of inhabitants "who are incessantly occupied with the necessities of the present hour and amongst whom no traditions preserve the memory of the past" may be valid in Canada, at least in the barren lands of the north.

Another factor worth considering is the possibility that the Indians may have had difficulty communicating their knowledge of cycles to each other. As David Thompson, writing at the end of the eighteenth century (published 1971), noted:

> The Indian forms his numbers of individuals, and appears to have no idea of numbers independent of them. Perhaps formerly the uneducated Shepherds, and Herdsmen obtained their ideas of numbers in the same manner, and (I) have frequently been told of Shepherds who could not by numbers count the Sheep in his flock, but by his own way could quickly tell if there was one missing.
>
> The Indians of the plains count only by tens, and what is above two tens, they lay small sticks on the ground to show the number of tens they have to count and in describing the herds of Bisons or Deer, they express them by a great, great many, and the space they stand on; for numbers is to them an abstract idea, but space of ground to a certain extent they readily comprehend and the animals it may contain; for they do not appear to extend their faculties beyond what is visible and tangible.

This difficulty in counting and expressing abstract numerical concepts may also have been a problem for the Eskimos in the North American arctic tundra. E. Strickler (personal communication, 1976), working with Eskimos in the Resolute Bay area above the Arctic Circle, noted several interesting facts. When the Eskimos count, their numbers can be effectively translated as: one, two, more than two, less than five, five. Thus if one Eskimo wished to express his knowledge of a regular 4-year cycle, his description would define a cycle of less than 5 years, a description that would not convey a true sense of regularity. Another fact is that these Eskimos tend to think in halves: if one were offered a carving for $20 and requested a lower price, the price would drop to $10, and nothing in between. This might suggest a sense of exactness beyond what would appear in population cycles, thus preventing these nature-conscious people from responding to a regularity that appears to be nearly exact on a large scale, but might not appear to be as exact when viewed locally over a period of only a few cycles. It is also conceivable that short-term fluctuations might appear as chaos, a concept that has recently acquired a statistical formulation (Li and Yorke, 1975).

An additional factor to add to the confusion is that Eskimos apparently are not too concerned with zoological exactness when describing animals as small as lemmings (Marsden, 1964). Many Eskimos do not distinguish between the brown lemming (*Lemmus*) and the summer coat of the varying lemming (*Dicrostonyx*), and they often include them with other mice and voles; some Eskimos use one word to mean either ermine, marmot, or lemming, whereas others use one word for lemming and weasel. Even in areas where brown and varying lemmings are distinguished and where migrations are recognized, as on the Alaskan arctic slope, lemming cycles as such do not appear to be recognized; this is curious since there is an awareness that these animals are more abundant in some years than in others, and, at Barrow, Alaska, there is even a recognition that brown lemmings have violent fluctuations in numbers while varying lemmings do not attain a high density in the area (corroborated by Pitelka, 1973). However, it may simply be that the lack of direct importance of the lemmings in Eskimo economics has not encouraged detailed observation of patterns in lemming populations.

Data Sources and Treatment

In any consideration of the possible existence of cycles in animal populations, conclusions must ultimately be based on an evaluation of population indices. The problems involved in assessing the value of the available indices have been discussed in detail by Elton (1942) and Keith (1963), but a brief review will be useful in order to recognize possible biases and sources of error which might limit the ability of these indices to portray actual changes in population levels.

The most important source of data has been fur returns, and these will be used almost exclusively, because they provide exact numbers for examination. However, it is important to remember that these figures represent more than just population per se, as the incentive to collect skins will depend on vacillating socioeconomic conditions and the varying trapability of animals, as well as random factors affecting the predator (i.e., trapper) population. For example, there seems to be a tendency for a higher percentage of most species to be trapped when the population is high rather than low, so peak population magnitudes may be exaggerated (Keith, 1963). This is not only due to increased probability of capture in higher densities, but also to the fact that when animals are starving (e.g., foxes; Elton, 1942) they are more willing to enter traps for bait. Also, there is the economic problem of supply and demand, which makes certain furs more desirable than others at different times. The data selected for this study were chosen to minimize the latter problem as much as possible. The Hudson's Bay Company, which for some time held a virtual monopoly in furs over most of the nonarctic regions of Canada, made fairly steady payments to trappers despite a fluctuating market. Although the territory covered by the various posts changed somewhat over the years, most of these changes have been carefully documented (Elton, 1942; Elton and Nicholson, 1942b). The Moravian Missions (agents for the Society for the Furtherance of the Gospel) are the principal source for data in Labrador until 1924, when the Hudson's Bay Company took over the fur-trading operations at the Mission settlements. The territory covered by the Mission posts was fairly constant, and the prices offered for furs changed very little, to avoid chaotic accounts (Elton, 1942). Canadian provincial statistics are somewhat more questionable because they lack these corrective factors, and the same may be true for fox pelts brought in for bounty in Norway.

None of these problems would be expected to induce real periodicities in population series, and most workers agree that, although the numbers cannot be construed as being directly proportional to true population figures, they are representative of existing population levels (Keith, 1963). Besides, the collection of data was unbiased by scientific theory (Elton, 1942).

The other major source of population data is regional hunting-kill estimates. These estimates are usually based on an extrapolation from incomplete reports, or postseason questionnaires, and potentially large discrepancies are inherent in the observations. Hunting regulations and weather can seriously affect the kill. It is therefore wise to view these data with a healthy skepticism and consider them as indicative at best.

3 *Do Cycles Exist?*

A great deal of effort has been made to construct mathematical models to explain wildlife oscillations but little has been done to clarify whether these oscillations can be termed periodic in the mathematical sense, or whether they can be considered as series of random variations. This is partially due to the tendency in the second quarter of this century to accept "periodicities" without a critical view toward what constitutes a cycle, so that a good deal of research was undertaken concerning causal factors for cycles before they were proved to exist (Butler, 1953; Hutchinson, 1975). When Cole (1951) injected some long-overdue skepticism into the subject of cyclic phenomena by suggesting that a series of random numbers, analyzed visually, might yield results little different from those biologists were discussing, most biologists were led to believe that all short-term oscillations are random, and that long-term oscillations may also be random (e.g., see *Cold Spring Harbor Symposium in Quantitative Biology*, Vol. 22, 1957).

However, computer technology makes feasible the use of two powerful and interrelated statistical tools to test a time series for genuine periodicity: autocorrelation and spectral analysis. The *autocorrelation function*, which was first applied to small mammal populations by Moran (1949a et seq.), is a measure of the degree of correlation between a series as observed and that same series if initiated after a specified time lag. If a series is periodic, a large positive correlation will be observed when the maxima of the series correspond exactly, and a large negative correlation will result when the maxima of one series correspond with the minima of the other. For a regular periodic phenomenon, the autocorrelation function will oscillate at regular intervals, the interval being determined by the frequency of oscillation of the original data. If the observations are oscillatory, whether regularly periodic or not, there should be at least one negative autocorrelation coefficient of some order of significance, but no necessarily regular oscillation of the autocorrelation function. The autocorrelation function avoids Cole's problem of finding an objective definition for a peak by considering the time series as a whole rather than as individual fluctuations.

The *power spectrum*, which is mathematically related to the autocorrelation function, analyzes the variance of a time series by partitioning the variance of the series of observations about its mean among frequencies that are harmonics of the length of the data set. Each time series is considered as a single realization of a stochastic process. The spectrum reveals to what extent a series is tuned to fundamental rhythms and clearly expresses relative contributions from different frequencies. Biologically, one would not necessarily expect to find finely tuned cycles, but if a series is fairly tuned to a relatively narrow range of frequencies, one can assert that it is cyclic. The spectrum can be considered as a weighting factor for making this decision. For general readable reviews of time-series analysis, Platt and Denman (1975) and Chatfield (1975) are particularly helpful. For more detail, consider Bloomfield (1976), Box and Jenkins (1970), and Jenkins and Watts (1968).

The statistical requirements of time-series analysis require normalized data, and it is useful to transform the data to logarithms to more nearly approximate a normal distribution (Moran, 1953a). Also, marked disparities between the natures of peaks and troughs, and other nonsinusoidal behaviors, can affect the analysis, and these disparities are reduced by the logarithmic transformation (Bloomfield, 1976). Logarithms will have an additional advantage in the discussion of the statistical mechanics of population dynamics. For these reasons all data in this study are subjected to a logarithmic transformation.

If the reader does not wish to study the formalisms of spectral analysis, the graphical output of spectral analysis (correlograms and spectra) has the significant advantage of giving easily comprehensible visual results, and the reader could therefore skip to the results. However, since time-series analysis is becoming more important in ecological studies, study of the details of this type of approach may prove useful.

The autocovariance of a variable z_t at time lag k is defined as the covariance between z_t measured at time t and z measured k time units later. The sample autocovariance function estimate for lag k is defined as

$$c_k = \frac{1}{N} \sum_{t=1}^{N-k} (z_t - \bar{z})(z_{t+k} - \bar{z}) \qquad (k = 0, 1, 2, \ldots, N-1)$$

where the mean, $\bar{z}$, is given by

$$\bar{z} = \frac{1}{N} \sum_{t=1}^{N} z_t$$

The estimate of the autocorrelation function, r_k, is

$$r_k = \frac{c_k}{c_0}$$

For a white noise process the expected value of the autocorrelation function is 1 for $k=0$ and 0 for $k \neq 0$. Since for N sufficiently large the expectation value of the autocovariance is distributed normally with mean zero and variance $1/N$ (Jenkins and Watts, 1968), a suitable confidence interval for r_k is

$$r_k \pm \sqrt{\frac{1}{N}}\, t_\alpha$$

where t_α is the estimate of Student's t with $N-1$ degrees of freedom for a two-tailed confidence band of width $(1-\alpha)$.

If the time series contains a trend, the values of r_k will not dampen to zero except for large k. The trend may sometimes dominate other features in this circumstance and should therefore be removed. Removal of a linear trend can be seen as possibly relating to some large-scale population movement, but the biological significance of removing higher-order trends (e.g., Bulmer, 1974) should be explicated before this direction is pursued. Measuring differences around a regression line instead of the mean is an alternative approach. Differencing, that is, forming a new series x_t such that

$$x_t = z_{t+1} - z_t$$

is often used in economic analyses (e.g., Box and Jenkins, 1970) but has several limitations. One of these was discussed in a study by Slutsky (see Chatfield, 1975), which showed that when averaging and differencing procedures are both applied to a completely random series, sinusoidal variation of the data can be induced. Nonetheless, the difference filter should not affect any inherent periodicities and can be a useful alternative.

As an example of the general appearance of an autocorrelogram for a periodic small mammal series, the classic Mackenzie River lynx series is shown in fig. 3. The autocorrelogram, first calculated by Moran (1953a), is oscillatory and damps out very slowly. The arrows indicate 95 percent confidence levels for r_1 (confidence levels for larger lags would have smaller values, and these estimates can thus be considered as maximum values for the entire series), and lags as large as $k=75$ to 80 are still significant. The period of the correlogram appears to be slightly less than 10 years. See the appendix for further examples.

Box and Jenkins (1970) demonstrate that large covariances can exist between neighboring values of the autocorrelation function, which can result in large periodicities which persist in the sample autocorrelation function when the theoretical value is expected to be close to zero. This can distort the shape of the autocorrelation function. It is best to avoid

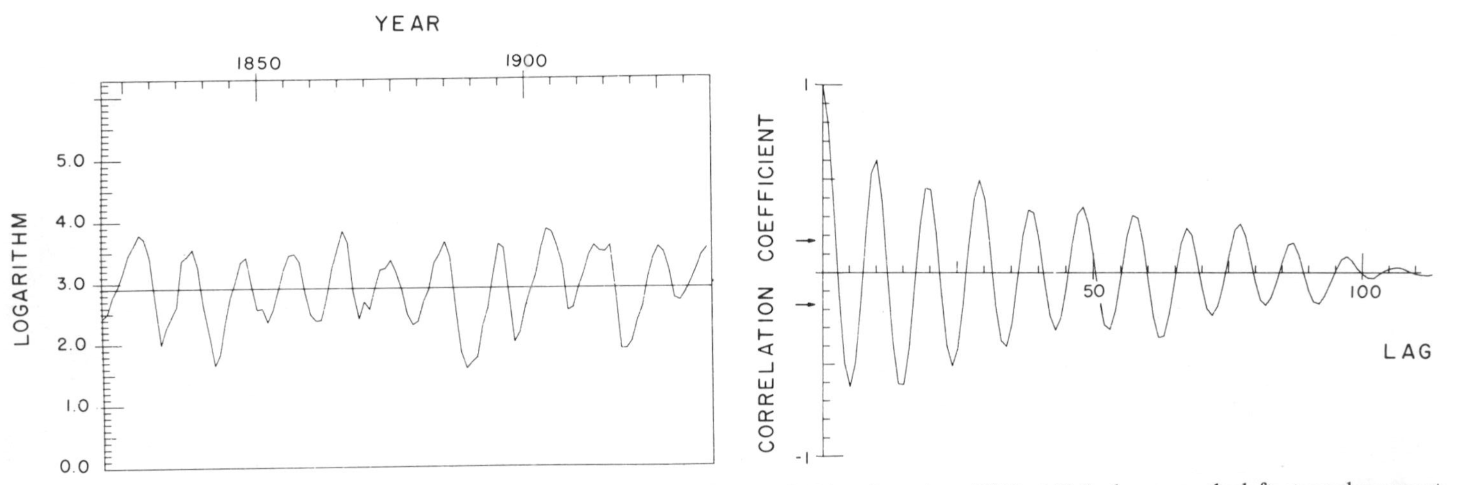

Figure 3. Logarithms of the population numbers from the original Mackenzie River lynx data (1821–1934), shown on the left, strongly suggest an oscillation, and this suggestion is supported by the oscillating autocorrelation function on the right. The autocorrelation has its first nonzero maximum at $t = 10$ years, but the gradual shift away from peaks at decades indicates that the actual period average is about 9.6 years. Arrows here, and in subsequent autocorrelograms, indicate a 95 percent confidence level for r_1.

making erroneous judgments by being a bit skeptical of the visual appearance of the autocorrelation function and treating the function as a guide for spectral analysis, since any periodicity in the autocorrelation function will only appear as a peak in the spectrum if the amplitude of the damped sine wave is large enough (Jenkins and Watts, 1968).

Fourier analysis demonstrates that a finite time series can be represented by a sum of sine and cosine terms of various frequencies. The periodogram represents least-squares fit of a finite number of sine and cosine functions of various frequencies to the time series z_t such that

$$z_t = \alpha_0 + \sum_{i=1}^{q} (\alpha_i \cos 2\pi\Delta f_i t + \beta \sin 2\pi\Delta f_i t) + \varepsilon_t$$

where the error term ε_t is minimized, $q = N/2$, f_i is the ith harmonic of the fundamental frequency $1/N\Delta$, and Δ is the sampling interval (Platt and Denman, 1975). If least-squares coefficients α_0 and (α_i, β_i) are a_0 and (a_i, b_i), respectively,

$$I(f_i) = \frac{1}{2} N (a_i^2 + b_i^2) \qquad (i = 1, 2, \ldots, q-1)$$

and

$$I(f_q) = I\left(\frac{0.5}{\Delta}\right) = N a_q^2$$

where $I(f_i)$ represents the periodogram intensity at frequency i. Thus the periodogram consists of $q = N/2$ uniformly spaced frequencies ranging from $f_q = 1/(2N\Delta)$ to $f_q = 1/(2\Delta)$. Bartlett (1954) calculated the periodogram for the Mackenzie lynx series (smoothed with the "Bartlett window," which will be discussed shortly), and his values are shown in fig. 4. A sharp peak, essentially confined to a narrow component centered around $f = \frac{12}{114}$ (period about 9.5 years), is obvious.

If the periodogram is allowed to be a continuous function defined at all frequencies $0 \leq f \leq f_q$, we have the sample spectrum

$$S_{xx}(f) = \frac{1}{2} N (a_f^2 + b_f^2)$$

It is of fundamental importance that the sample spectrum can be shown to be the Fourier transform of the estimated autocovariance function:

$$S_{xx}(f) = 2\Delta\left[c_0 + 2\sum_{k=1}^{N-1} c_k \cos 2\pi\Delta f k\right] \qquad 0 \leq f \leq \frac{1}{2\Delta}$$

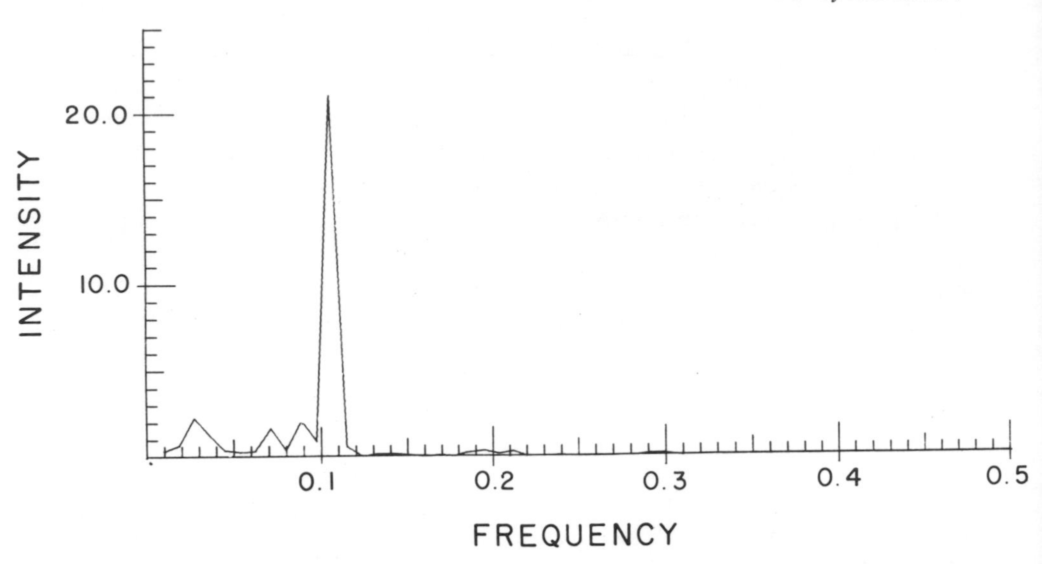

Figure 4. Periodogram analysis for the Mackenzie River lynx data demonstrates a clear peak for a period of slightly less than 10 years (frequency = 1 per period).

The expected value of the sample spectrum for N large is the true power spectrum:

$$\Gamma_{xx}(f) = \lim_{N \to \infty} E[S_{xx}(f)]$$

For a white noise process in discrete time (Jenkins and Watts, 1968) the expectation value of the autocovariance function is σ_z^2 (the variance of the series z_t) at lag zero, and zero everywhere else; from this it can be shown that

$$\Gamma_{xx}(f) = \sigma_z^2 \Delta \quad -\frac{1}{2\Delta} \leq f < \frac{1}{2\Delta}$$

Thus in the range $-1/(2\Delta) \leq f < 1/(2\Delta)$, all frequencies contribute the same amount of power for a white noise process.

Because the spectrum is the Fourier cosine transform of the autocovariance function, knowledge of the autocovariance function and the power spectrum are mathematically equivalent. The two are complementary visually since autocorrelation (derived from the autocovariance function) represents analysis in the time domain, whereas the power spectrum represents analysis in the frequency domain. The correlogram provides information concerning the relationships between values of a series which are separated in time; the spectrum reveals the extent to which a series is tuned in fundamental rhythms, allowing one to see the relative contributions of different frequencies. The spectrum might be viewed as a weighting factor arguing for or against periodicity in a given series.

Neither the periodogram (Hannan, 1960) nor the spectral estimator $C_{zz}(f)$ (Jenkins and Watts, 1968) is a consistent estimator, since in both cases as N increases the variance approaches a constant value rather than zero, so that both estimators are subject to violent fluctuations about their theoretical values. To reduce this variance, it is necessary to smooth the sample spectrum. This is done by splitting the original series into M subseries of length N/M and calculating the sample spectrum of each subseries at each frequency. The mean for all M subseries at each frequency gives the smoothed spectral estimate (C_{zz}) at that frequency. This is mathematically equivalent to calculating the spectral estimate by weighting the autocorrelation estimates with a series of weights λ_k such that

$$C_{zz}(f) = 2\Delta\left(c_0 + 2\sum_{k=1}^{N-1} \lambda_k c_k \cos 2\pi f k \Delta\right) \qquad 0 \le f \le \frac{1}{2}$$

Intuitively, this seems a reasonable process since as k increases, the precision of c_k decreases (because the series on which the calculation is based becomes shorter) and λ_k is designed to give less weight to values of c_k as k increases. The set of λ_k is called a *spectral window*; the "width" of the spectral window is determined by the length of the M subseries. In a sense the spectral window plays the role of a noise filter: the wider the window, the more noise passes through. But caution must be exercised to avoid making a window too narrow because in order to separate two frequencies at f_1 and f_2, the window width must be of the order $1/(f_1 - f_2)$.

Several windows have been used in studies of small mammals. The Bartlett window with $\lambda_k = 1 - k/M$ for $k = 0, 1, \ldots, M$ (Bartlett, 1950, 1954; Bulmer, 1974) is seldom used because weights λ_k decay linearly instead of remaining close to unity for small k (Bloomfield, 1976). The Tukey window (Finerty, 1972) and the Parzen window (Williamson, 1975) will give nearly the same estimate for a given time series (Chatfield, 1975; Jenkins and Watts, 1968). The Tukey window with

$$\lambda_k = \frac{1}{2}\left(1 + \cos\frac{\pi k}{M}\right) \qquad (k = 0, 1, \ldots, M)$$

will be used in this study. The shape of this window is shown in fig. 5.

The length M of the subseries is inversely proportional to what is called the *bandwidth* of the window. The narrower the bandwidth, that is, the

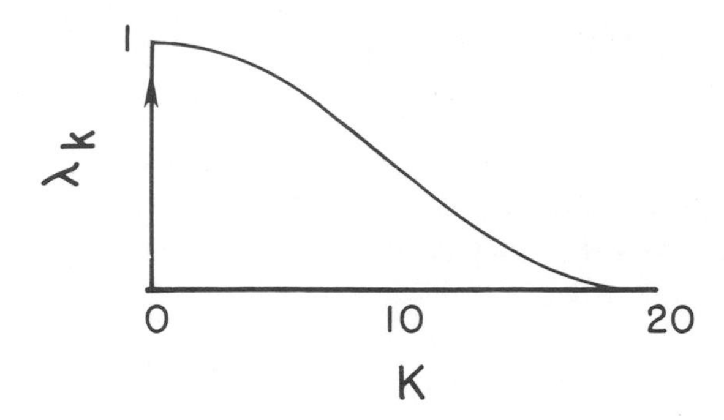

Figure 5. Shape of the Tukey "window" for $M = 20$. As the value of k increases along the horizontal axis, the weighting factor λ_k decreases, since the precision of the estimate decreases.

larger the M, the larger the variance of the spectral estimator. However, decreasing M increases the bias (the difference between the theoretical spectrum and the spectral estimate). Too small a value for M may smooth out important features of the spectrum, and too large an M will result in erratic variation. For an optimal spectrum it is desirable to have the highest fidelity (smallest bias) and stability (smallest variance). As each of these can only be achieved at the expense of the other, a compromise is necessary. The effects of the compromise can be observed by using a technique known as *window closing*. A spectrum is estimated with a wide bandwidth, providing high stability but poor fidelity. The bandwidth is gradually narrowed if more detail can be revealed without producing spurious peaks (peaks that are not suggested by wider bandwidths).

For the Tukey window, the bandwidth (window width) is given by

$$b = \frac{1.33}{M}$$

and there are $2.67N/M$ degrees of freedom (Jenkins and Watts, 1968).

The suggested lag M lies in the range $N/10 \leq M \leq N/4$ (Platt and Denman, 1975). This study usually employs three window widths corresponding to $M = N/10$, $\frac{1}{2}(N/10 + N/4)$, and $N/4$. Each of these spectra can be examined to see if it converges toward the others without becoming unstable. Jenkins and Watts (1968) explain in detail that for v degrees of freedom

$$\frac{vC_{zz}(f)}{\Gamma_{zz}(f)}$$

is distributed as a χ^2, so that a $(100 - \alpha)$ percent confidence band for

$\Gamma_{xx}(f)$ is

$$\frac{vC_{zz}(f)}{\chi_v(1-\alpha/2)}, \frac{vC_{zz}(f)}{\chi_v(\alpha/2)}$$

where the probability that

$$\chi_v^2 \leq \chi_v\left(\frac{\alpha}{2}\right)$$

is $\alpha/2$. This confidence interval changes with each frequency. By plotting

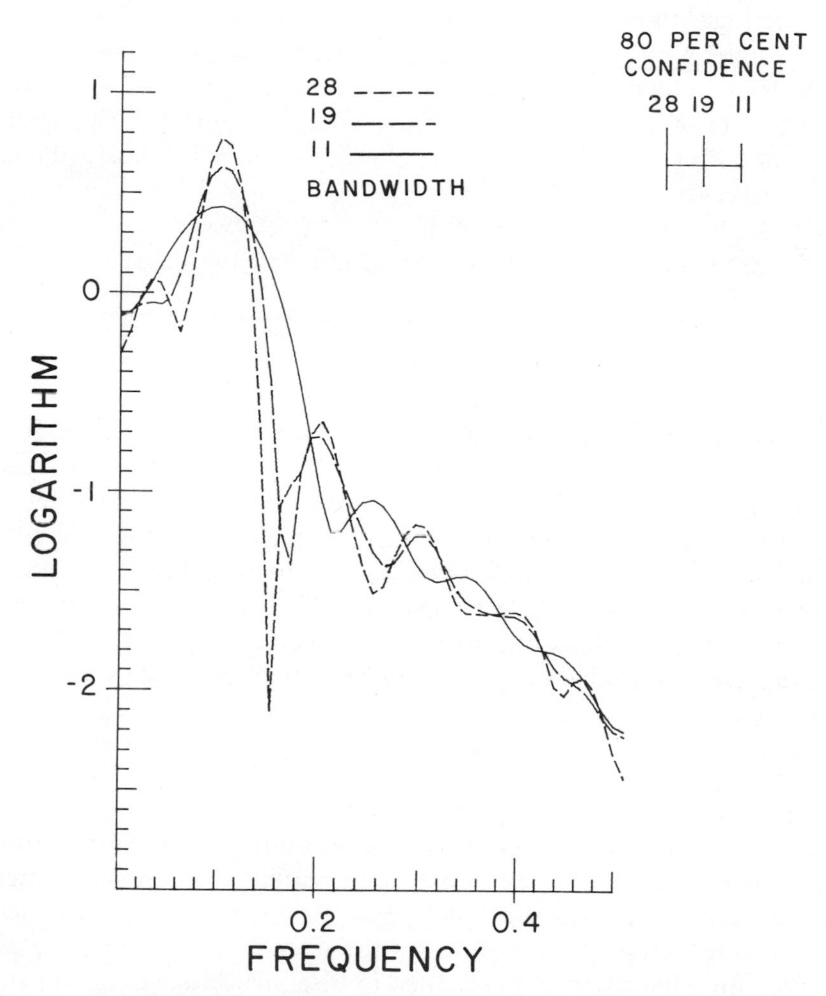

Figure 6. Logarithm of the smoothed autospectrum versus frequency for the Mackenzie River lynx data, 1821–1934. The strongly emergent peak near a 10-year period (frequency = 0.1) is obvious.

logarithms of the spectral estimates, the interval becomes

$$\log C_{zz}(f) + \log \frac{\nu}{\chi_\nu(1-\alpha/2)} \qquad \log C_{zz}(f) + \log \frac{\nu}{\chi_\nu(\alpha/2)}$$

and the confidence band can be represented by a constant interval on the vertical axis. The logarithmic scale also helps to reduce wide variations in the spectrum and depicts proportional changes in power, which are of prime interest when testing for evidence of periodicity and not goodness of fit.

When examining spectra it is important to note whether the plot is linear or logarithmic. A linear plot (e.g., the Bartlett lynx periodogram, fig. 4) is quite a bit more dramatic than a logarithmic plot, but although the understated nature of the latter may seem less convincing, it can be easier to interpret. The logarithm of the smoothed autospectrum for the Mackenzie River lynx data is shown in fig. 6, with a clearly emergent peak near $f = 0.1$ (period of 10 years).

With these mathematical techniques it will be possible to make more objective statements about the existence of population cycles.

Cycles and Spectra: One Approach

The usefulness of time-series analysis of small mammal data is evident in a series of papers by Bulmer (1974, 1975a, 1975b, 1976) in which the author intelligently approaches a previously unexplored cache of data derived from the Canadian Hudson's Bay Company records (from Jones, 1914). Using autocorrelation and spectral analysis (with a Bartlett window), Bulmer (1974) examined 10-year cycles in Canada for the eighteenth, nineteenth, and twentieth centuries. Those species which on the basis of these calculations evidenced periodicity were then analyzed by fitting to a "mixed model":

$$x_t = \mu + a \sin 2\pi\omega(t - \phi) + Bx_{t-1} + e_t$$

where x_t is the logarithm to the base 10 of the population size (i.e., number of animals trapped) in biological year t, a the amplitude of the oscillation, $1/\omega$ the period in years, ϕ the phase lag in years, μ the population mean, and e_t an error term. The sine function expresses the regular periodicity, and the autoregressive term expresses the irregularities of amplitude by incorporating the error term (assumed to be a measure of environmental factors) into the process. Bulmer points out that the autoregressive term is also biologically reasonable because whatever factors are producing oscillations would be expected to affect birth and death rates, and the population for a given year would depend on the population the year before as well as the birth and death rates. This model would be expected to produce spectra with a single narrow peak (due to the sine function) followed by a decline with increasing frequency (due to the autoregressive term). The spectrum (= periodogram) for the lynx data (Bartlett, 1954; data from Moran, 1953a) was shown in fig. 4, and is what one would expect for this model. A mean-square analysis for fit of the lynx data to this model using $1/\omega = 9.63$ years (the period length estimated by Elton and Nicholson, 1942b) provides a satisfactory fit.

Following Bulmer's approach, periodograms for 11 species (data from Jones, 1914; snowshoe hare from MacLulich, 1957) for the period 1848–1909 were constructed (fig. 7); no trends were removed before calculating periodograms. Bulmer stated that "in all cases where any peak was obvious it was confined to a single component which was the same for all these species and which corresponded to a period slightly less than 10 years. The data were, therefore, analyzed by the method just described, taking the period as 9.63 years and removing a linear or quadratic trend if necessary" (Bulmer, 1974; p. 704). Bulmer found significant evidence for 10-year periodicity in the following species: fisher (*Martes pennantes*), red fox (*Vulpes vulpes*), lynx (*Lynx canadensis*), marten (*Martes americana*), mink (*Mustela vison*), muskrat (*Ondatra zibethicus*), skunk (*Mephitis mephitis*), wolf and mostly coyote (*Canis lupus* and *Canis latrans*, years 1879–1909), wolverine (*Gulo gulo*), and snowshoe hare (*Lepus americanus*). Examination of the periodograms here supports Bulmer's contentions for lynx, coyote, and color phases of the red fox. Periodograms for other species show evidence for low-frequency trends (muskrat, mink, skunk, and wolverine) or large secondary peaks (fisher, marten, and red phase of the red fox). While lynx and fox do peak at a frequency corresponding to a period of around 9.63 years, this is not clear for the other species. It is possible that detrending *before* calculating periodograms might have been used by Bulmer to correct this, but this is unclear.

Although this approach is quite useful for reemphasizing the regular periodicities in several mammal populations, there are several problematic facets that are bothersome. The biological significance of a period of 9.63 years is not immediately obvious, and the fact that the data are measured in annual units suggests that we can accurately state only that the period is somewhere in the vicinity of 9 or 10 years. It is not clear at this point whether or not the variance of the period is large enough to require explanation. Also, assuming that the period length is the same for all species may be misleading, since we cannot yet definitively assert that the

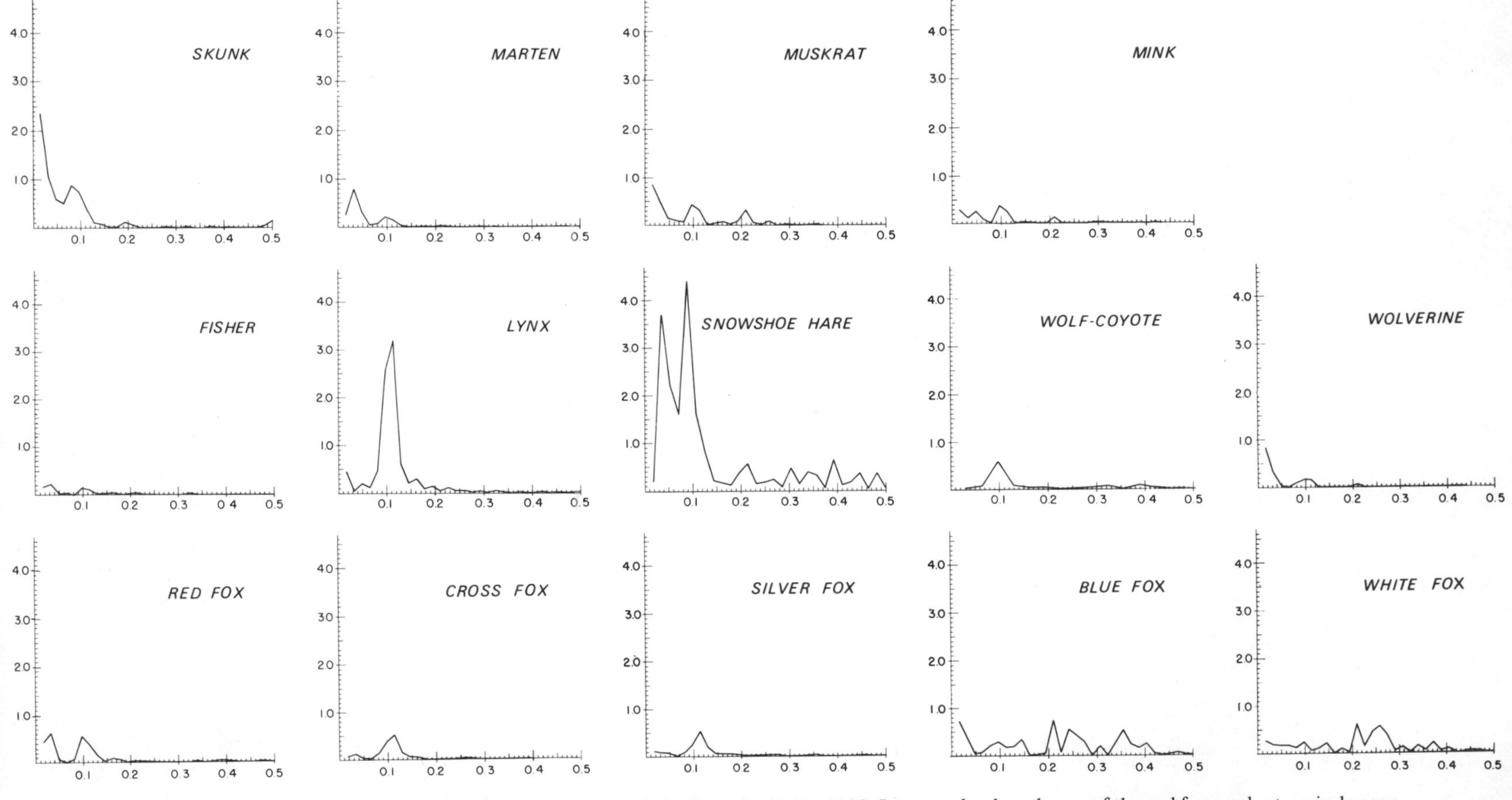

Figure 7. Periodograms for 14 species of small mammals in Canada, 1848–1909. Lynx and color phases of the red fox peak at periods near 9.63 years, but peaks in this vicinity for other species are not as clear as Bulmer (1974 et seq.) has suggested. Bulmer's smoothing techniques may explain his clearer results, but what these smoothing techniques represent biologically is somewhat uncertain.

period length is the same in all areas for any given species. However, one can still without hesitation conclude, as Bulmer did, that the species mentioned were cyclic in the last half of the nineteenth century. Similarly, analyzing other Hudson's Bay Company data sets, Bulmer demonstrated probable cyclicity in fisher, lynx, mink, and wolverine for the period 1920–69; and in lynx, marten, mink, and possibly wolverine for the period 1751–1847 (with some gaps in the data). Agricultural and urban invasion may account for changes in population flux in the twentieth century, as they have clearly changed the distributions of some species (Petersen, 1966). Unreliability of some of the eighteenth- and early nineteenth-century data, as Bulmer noted, may account for absence of cycles in certain species during that period. For the nineteenth century it seems quite possible that massive clearing of the forests opened new avenues of dispersal and new marshy habitats along streams, thus enhancing the phenomenon of emigration from areas where cycles were endemic (F. A. Pitelka, personal communication) and making the existence of cycles more obvious.

Evidence for Four-Year Cycles

Lemmus and Dicrostonyx

Among rodents said to exhibit cyclic population fluctuations, lemmings, muskrats, and voles are generally mentioned. In this study emphasis will be on lemmings and muskrats, and their predators; further evidence for other species is available in Elton's (1942) classic *Voles, Mice and Lemmings*, and in an extensive review by Krebs and Myers (1974).

Two genera of lemmings are found in the arctic tundra, *Lemmus*, which is brown year-round, and *Dicrostonyx*, which is grayish in the summer and white in the winter. *Lemmus* as a group is characterized by the primitive condition of its lower incisors, which are extremely short, and the absence of a seasonal pelagic dimorphism (Ognev, 1963). It has been suggested that members of the genus *Lemmus* are probably more specialized than those of any other related genera (Miller, 1912).

The Norwegian lemming, *Lemmus lemmus* (Linnaeus), is distributed over about one third of Norway, including nearly all of the mountain plateaux, and is the only form of *Lemmus* in Europe west of Russia (fig. 8). The preferred home of the Norwegian lemming is the subalpine zone between the upper limit of the silver birch forest and the true alpine barrens running vertically past the permanent snow line, where the

Lemmus species. *L. trimucronatus* left

Hudson Bay lemming, *Dicrostonyx hudsonius*

Figure 8. Distribution of *Lemmus* in Scandinavia. (Based on van den Brink, 1967.)

ground is marshy and tussock-covered, with dwarf birch trees and shrubs, mosses, sedges, grasses, and cloudberry plants. Norwegian lemmings live chiefly on the mainland and several islands in the north, and are only found in lowland areas in migratory years. These Norwegian mountain lemmings do not form a continuous connected population, but live in mountain blocks separated by valleys that dip into woodland zones (Elton, 1942). In normal years the lemming's life is secluded. In some years numbers may increase enough to push individuals into the upper parts of wooded areas, but a migration does not necessarily result (Collett, 1911–12).

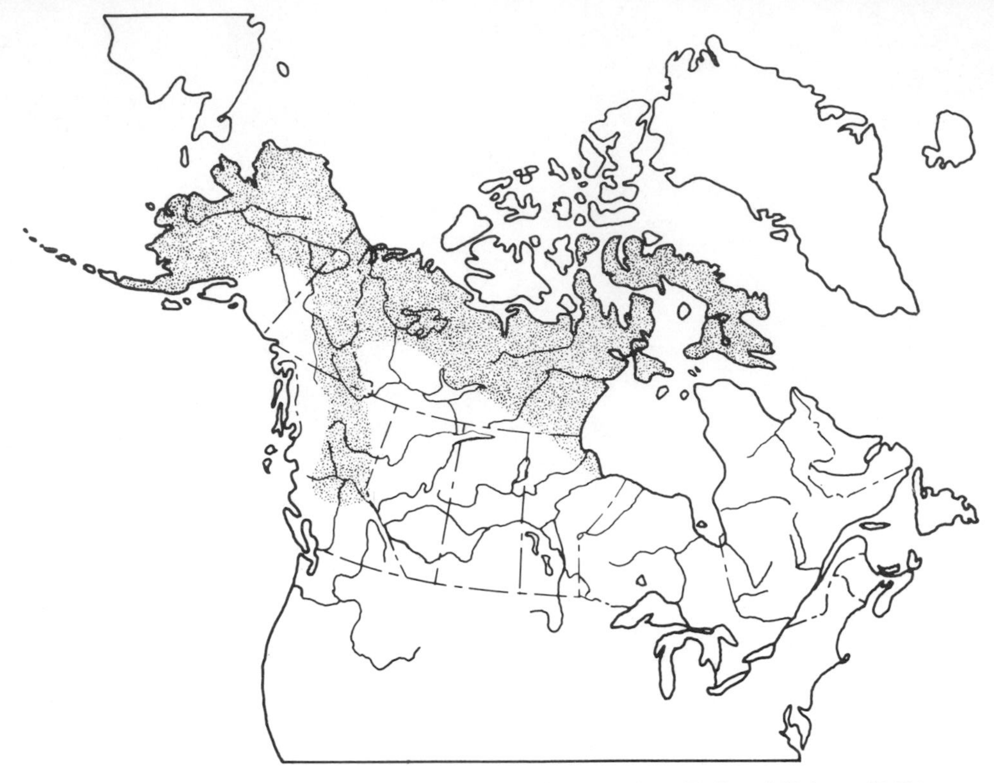

Figure 9. Distribution of *Lemmus* in Canada. (Based on Hall and Kelson, 1959.)

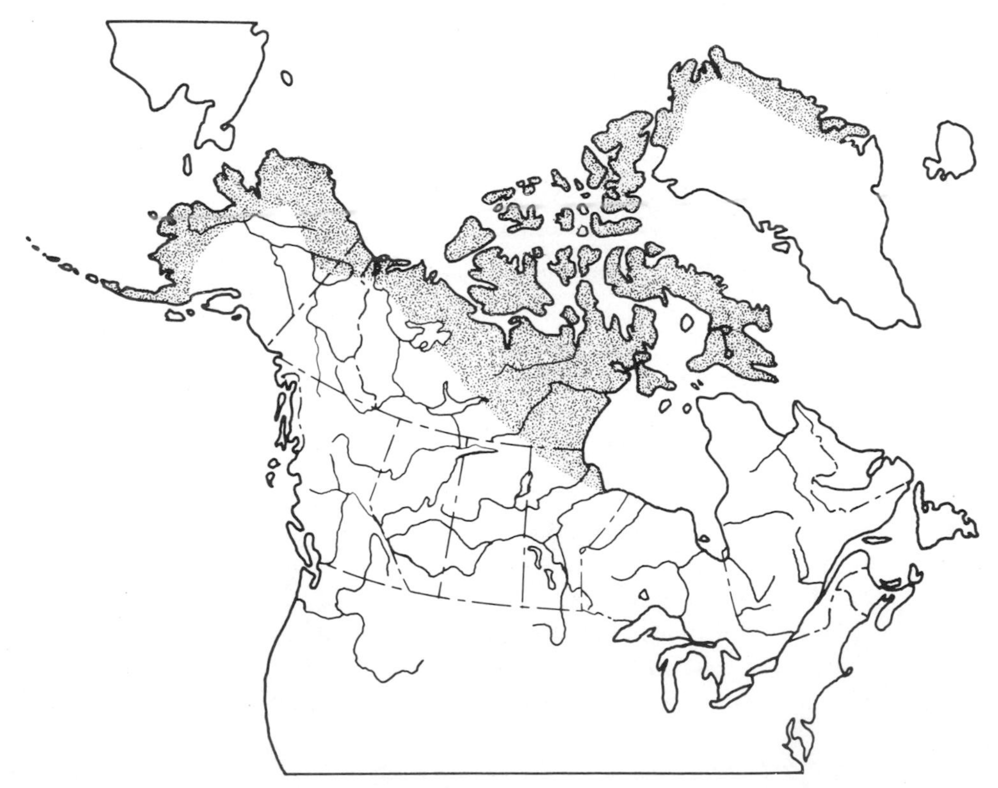

Figure 10. Distribution of *Dicrostonyx groenlandicus* in Canada. (Based on Hall and Kelson, 1959.)

Lemmings feed on grasses in summer and on low mosses under the heavy snows in winter, which is probably a safety advantage for the lemmings because without a protective brush cover, they would be exposed to severe predation were they to come out onto the snow to feed (Marsden, 1964).

In Canada, both *Lemmus* (commonly *L. trimucronatus* Richardson or brown lemming) and *Dicrostonyx* [mostly *D. groenlandicus* (Pallas) and *D. hudsonius* (Pallas) or varying lemming] abound, inhabiting the vast tundra and barren lands (figs. 9 and 10). There are several so-called subspecies of *L. trimucronatus* in Canada, but all the mainland forms are a single geographically varying species (Davis, 1944), which is closely related to the Scandinavian species (Banfield, 1974). However, *L. trimucronatus* differs ecologically from *L. lemmus* in several ways: (1) although the southern portion of the Canadian species' range reaches into the alpine tundra, *L. trimucronatus* primarily inhabits coastal plains where grasses and sedges are dominant on wet lowlands and polygonized ground and where shrubs and forbs are dominant on ridges and hummocks (Batzli, 1975); and (2) although a seasonal spring migration to higher land to escape spring flooding often occurs, no long emigrations have been recorded in this species (Batzli, 1975).

Like *L. trimucronatus*, *D. groenlandicus* is also divided into a large number of conspecific geographical races, but *D. hudsonius* (fig. 11), which is the only lemming living on the Atlantic coast of Labrador and the east coast of Hudson Bay, is a distinct species, separated sharply from *D. groenlandicus* by tooth characters (Anderson and Rand, 1945). *Dicrostonyx* differs from *Lemmus* by having smaller ears, highly developed claws (the powerful third and fourth claws being enlarged in winter due to growth of the bulbous portion underneath), and distinctly more general-

Figure 11. Distribution of *Dicrostonyx hudsonius* in the Ungava peninsula. (After Banfield, 1974.)

ized skull and molars (Miller, 1912). It is also distinguished by seasonal color variation, ranging from winter white to a summer pelage which varies greatly with subspecies (Anderson and Rand, 1945; Ognev, 1963). Grasses are the common food source in the summer, and in winter the lemmings work under the snow eating the basal portions of grasses and sedges and leaving the dry tops on the surface as litter (Thompson, 1955a; Marsden, 1964). In winters of extreme need they may eat the bark of dwarf birch and mountain ash (Collett, 1911). When winter use of vegetation is extensive, it is more conspicuous on the coastal plains than it is in interior portions of the tundra (Pitelka, 1957). Collett (1911–12) characterized lemmings as being extremely greedy, eating so energetically that their bodies will begin to shake, and Elton (1942) characterized them as "fat, busy, agile mowing machines"!

During the winter the lemming grows a long fur coat over its wooly underfur and burrows under the snow for protection against the cold. In winter months lemmings will actually avoid places with poor snow cover (Kalela et al., 1961) because the subnivean temperature is proportional to the ambient temperature and the depth of the snow (MacLean et al., 1974). In some areas in the autumn the soil and air temperatures can become unfavorably cold before an insulating blanket of snow is present, leaving only hollows where the first early snow has accumulated as favorable microhabitats for lemmings (Fuller, 1967); in other areas (e.g., Barrow, Alaska; MacLean et al., 1974) the subnivean environment does not have as much spatial variation in temperature. Under the snow the lemming builds a nest where it is safe from most predators except for weasel, ermine, and arctic fox, which will dig into the snow. The frequency with which a lemming comes to the surface will depend on the depth of the snow, especially on the continuity of the snow cover, as well as on the population density (Pitelka, 1957).

As the snow begins to thaw in the spring there is danger of flooding in the burrows, and in some years lemmings drown by the thousands (Marsden, 1964). Changes in the snow also force these rodents out of their nests because of the imminent danger of collapse. Norwegian lemmings seem to avoid these problems by moving to the peatlands of the alpine zone and parts of the forest zone which are on higher ground, returning to the lower areas and the protection of the snowbanks for winter (Kalela et al., 1961). During the summer the lemmings burrow into the ground and build new nests. Among Canadian lemmings, *Dicrostonyx* prefers well-drained, fairly dry, and excavatable ground, usually ridges and rolling uplands (Pitelka, 1957) where dietary requirements (dicot leaves) and the need for dry burrow sites can be met (Batzli, 1975); whereas *Lemmus* prefers wet places and the deeper soil of lower areas. There are exceptions to this dictum and it may only be a clear distinction when both species are present in abundance. Just as Canadian *Lemmus* inhabits low-lying places and moves to higher ground in late spring to escape flooding, *Dicrostonyx* occasionally undertakes a shift of habitat also (Marsden, 1964).

Summer is the height of the breeding season for lemmings. Breeding begins when the snow melts in the spring, and it is possible for a litter to be produced nearly every 3 weeks (Krebs, 1964a). The gestation period is approximately 21 days for both *Lemmus* and *Dicrostonyx*, with an average litter size of four to nine (Collett, 1911–12; Burton, 1962; Banfield, 1974). Female lemmings are often able to conceive by the twenty-seventh day after birth, and males mature in about 30 days (Marsden, 1964). Breeding does occur in winter, but not at all times (Krebs, 1964a; Marsden, 1964). In the wild it is rare for either *Lemmus* or *Dicrostonyx* to live more than 1 year (Ognev, 1963; Banfield, 1974).

The great reproductive capacity of lemmings provides food for many predators. In Norway, Collett (1911–12) recognized the rough-legged hawk (*Archibuteo lagopus*), ground owl (*Asio accipitrinus*), snowy owl (*Nictea scandiaca*), ravens, crows, and several falcons as avian predators; and arctic fox (*Alopex lagopus*), red fox (*Vulpes vulpes*), ermine (*Mustela erminea*), and weasel (*Mustela nivalis*) as major small mammal predators. Some of these predators, such as the snowy owl, are present only when lemmings are abundant since they are migratory and will only stop to nest if sufficient food is available. Some birds, such as the long-tailed jaeger (*Stercorarius longicauda*), will be attracted into the high fjelds far south of their usual range during lemming peaks. In the northern parts of Norway, Lapp owls (*Syrnium lapponicum*), parasitic (*St. parasiticus*) and long-tailed jaegers, and several kinds of sea gulls are added to the list of predators.

In eastern Canada, common lemming predators are the snowy owl, the short-eared owl (*Asio flammeus*), glaucous gull (*Larus hyperboreus*), raven, gyrfalcon (*Falco rusticolus*), rough-legged hawk, pomarine (*St. pomarinus*), and parasitic and long-tailed jaegers among avian predators; and least weasel (*M. rixosa*), ermine, arctic fox, red fox (*Vulpes fulva*), wolf (*Canis lupus*), wolverine (*Gulo gulo*), and grizzly bear (*Ursus americanus*) among mammals (Banfield, 1974). On the coastal tundra of Barrow, Alaska, the most significant predators are pomarine jaegers, snowy owls, and least weasels, with the arctic fox often appearing as a strong secondary predator (Pitelka, 1973).

While avian predation on lemmings is significant and must be included in any model of lemming flux, these predators will not be discussed in detail since there are essentially no long-term numerical data on any of these species to support any contentions concerning their cyclicity. Their breeding habits effectively determine their relation to lemmings, either staying to breed for migratory birds if food is abundant, or not breeding for residents. The major concern here will be small mammal predators ("small" being as large as wolves or wolverines), for which there are fur collection data and hunting statistics.

One of the major problems in population biology is that in many systems, it is far more difficult to collect actual data to verify or disprove theories than it is to generate theories (Hutchinson, 1975), and lemming populations are an excellent example of this. Although many studies on lemmings have been pursued in depth, few have persisted to produce a data base long enough to be useful for mathematical analyses. As a result, most of our knowledge of the regularity of lemming fluctuations is based on fur returns of lemming predators. Although these fur returns are unreliable indicators of absolute population changes, they do tend to reflect the observations of trappers and naturalists (Elton, 1942). The only direct evidence available is of two types. The first is records of lemming migrations in Norway, which, as has been mentioned, occur somewhere in the country each 3 to 4 years (Collett, 1898 in Wildhagen, 1952; Elton, 1942). The second is two data sets from the Canadian tundra. The first of these was published by Shelford (1943, 1945) and is a record of the abundance of *Dicrostonyx groenlandicus* from 1929 to 1943 in the Churchill area of northern Manitoba on the shores of Hudson Bay (fig. 12). The data set is rather short for meaningful time-series analysis, but there are clear spacings between peaks of 3 and 4 years. The other data set comes from the work at Barrow, Alaska (fig. 13), and shows population densities in *Lemmus trimucronatus* from 1946 to 1971. Here, again, spacings of 3 and 4 years are clear in the first half of the data set, but spacings after 1960 are somewhat irregular. This irregularity is especially significant since, for example, the peak in 1971 is thought to have been produced by an influx of migrants from a neighboring population and probably does not represent an increase in the local lemming population (Pitelka, 1973). This should emphasize the fact that numbers in themselves cannot tell anything without the biological observations necessary to clarify them.

As a point of reference, we will summarize a typical cycle in *Lemmus trimucronatus*, as viewed by Pitelka (1973) on the coastal tundra at Barrow, Alaska. A peak year includes a winter and the summer following. In the winter the lemmings mow the dormant perennial grasses at the ground surface, leaving a conspicuous layer of dead plant material on the

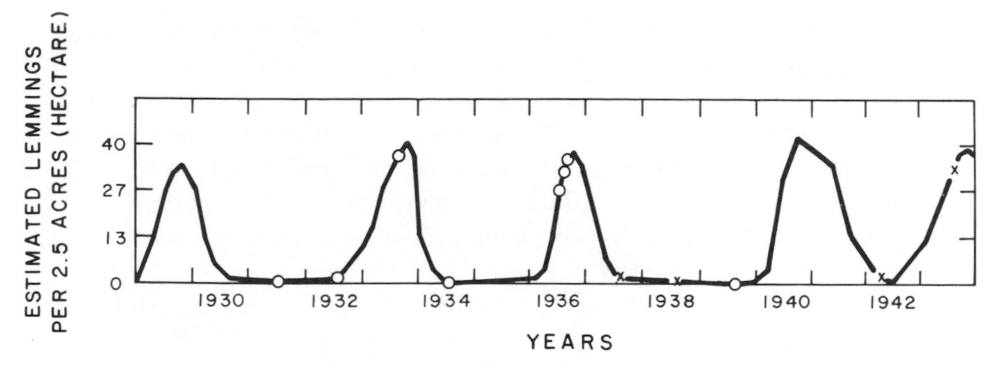

Figure 12. Shelford's (1943, 1945) lemming data for the Churchill area of northern Manitoba, Canada.

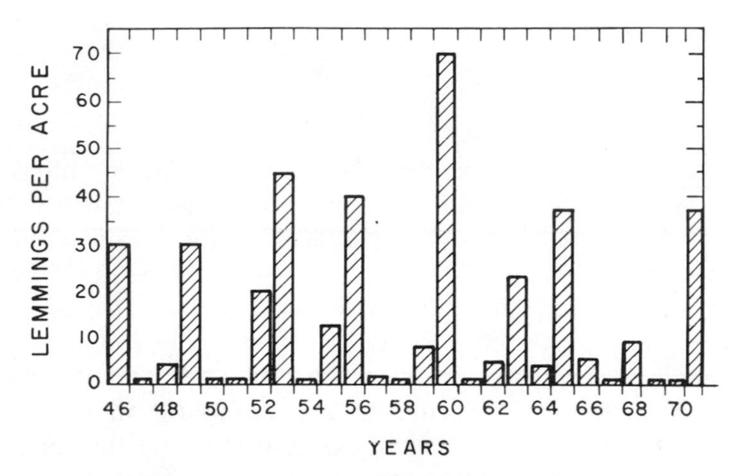

Figure 13. Population densities of the brown lemming, *Lemmus trimucronatus*, at Point Barrow, Alaska, in summer. (After Schultz, 1969; Pitelka, 1973; Krebs and Myers, 1974.)

ground surface and reducing standing vegetation to a level where lemmings are easily seen by predators. The rise in numbers begins in the prepeak year and continues through the winter. The increase toward a peak can become evident in late summer of the prepeak year, or during the winter of the prepeak year, and the chief source of the increase in numbers is survival and recruitment during that winter. Maximum density is reached in late spring. In years between peaks, brown lemmings occupy optimal habitats on polygonal ground, but in peak years they disperse throughout all available habitats.

Avian predation begins with exposure of the lemming population at the onset of snowmelt. Avian predators, whose numbers are highly variable during the prepeak summer, pace the increase of the lemmings at that time. Heavy avian predation begins in the peak summer and continues until the postpeak winter. Recruitment during that summer can slow, but not reverse, the decline. Predation by least weasels, and arctic foxes when they are present, can continue throughout the postpeak winter.

Several physiological phenomena are noted: lemmings attain highest body weights in the late winter and early summer of peak years; precocious breeding is most common among lemmings born in the prepeak summer; and reproductive rates decline during peak summers.

This description is generally similar to what is observed in *Lemmus* in Norway, with the major exception that the structure of the Norwegian landscape creates a situation in which dispersal of peak populations can result in massive migratory movements (Collett, 1911–12).

With this background in lemming fluctuations, the fluctuations in lemming predators may be more comprehensible.

Alopex lagopus

The arctic fox, *Alopex lagopus*, breeds in a range that encompasses the whole of the arctic tundra zone of North America (fig. 14) and Europe (fig. 15), plus areas of alpine tundra in Scandinavia. This is essentially a single species in both the Old and New World arctic. The arctic fox generally lives above the timberline or near the seacoast. In these areas the arctic fox feeds almost exclusively on lemmings. If the fox lives near the sea, he may rely heavily on marine invertebrates, fish, sea mammals, and nesting seabirds and their eggs during summer months; in winter he may wander out onto the sea ice and eat the remains of seals left by polar bears (Burton, 1962; Chesemore, 1968; MacPherson, 1969). Because the foxes live in a region of low biotic abundance, they are profoundly affected by any decline in the lemming population. The effects are only mitigated in maritime regions, where other food sources are abundant.

Arctic foxes are known to take spectacular migrations, sometimes wandering south for hundreds of miles (Petersen, 1966; Banfield, 1974), and these wanderings are usually associated with lemming population crashes. However, the first wave of migrants, which usually moves after the freeze-up about November, is primarily composed of well-fed males (Banfield, 1974), and this may indicate that foxes are taking advantage of good years to expand their habitats.

The fox normally whelps underground in a breeding den which is used for many years. Whelps are born in the late spring after a gestation period of 51 to 57 days, and are weaned in midsummer. In a given year only about one third of the 1- to 2-year-old vixens breed, whereas about five sixths of

Arctic fox, *Alopex lagopus*

the 3-year-and-over group breed; age is the only factor thus far known on which breeding depends (MacPherson, 1969). Since the number of surviving young per litter is sizably affected by the availability of food (i.e., lemmings), MacPherson (1969) hypothesized that delayed breeding might be an evolutionary advantage, since "most arctic foxes destined to survive their first winters are born in years of lemming abundance, and it is unlikely that the following year, or even the year after, will bring another such bumper season. Breeding in the first or second years of life may thus be non-advantageous . . . it is possible that early breeding . . . is actively selected against . . . owing to the risks and energy drain of parenthood being counterbalanced so rarely by the production of descendants" in years following a cyclic peak when lemming numbers are low. Even if arctic foxes breed during a period of lemming scarcity, prenatal mortality may increase and entire litters may be lost; and if some whelps are successfully raised at this time, the number weaned per litter may be half of that in a peak lemming year (MacPherson, 1969; Hoffmann, 1974).

The average litter size is between 10 and 11 pups and does not depend

Figure 14. Distribution of the arctic fox, *Alopex lagopus*, in Canada. (After Hall and Kelson, 1959.)

Figure 15. Distribution of the arctic fox, *Alopex lagopus*, in Scandinavia. (Based on van den Brink, 1967.)

on the age of the vixen. The farther north the fox lives, the larger the mean litter size seems to be, which may be the consequence of heightened seasonal contrast in food resources.

Two color phases can be recognized in the arctic fox, white and blue. The proportions of the two phases vary greatly from place to place. In the eastern Arctic (Baffin Island and coastal Labrador) the percentage of blues varies between 1.1 and 4.4 percent; in Greenland the ratio is approximately 50 : 40; in the Aleutian and Pribilof islands blue foxes predominate, and only blue foxes occur in Iceland (Banfield, 1974). The Danish biologist Braestrup (1941) suggested that the two colors indicated two races of arctic foxes, the white ones being "lemming foxes" having access to lemmings and the blue being "coastal foxes" without access to lemmings. Braestrup came to this conclusion because he observed that in the northern portion of western Greenland the proportion of white fox in the population varied from 4 percent to 60 percent in different years, and this increase was connected with migration. Most of the northern districts in Greenland had peak populations in 1923, 1927, 1931, and 1938, and the percentage of white foxes varied in the same way. The peaks in the percentage of white foxes decreased toward the southern regions and gradually disappeared. Braestrup connected this with the fact that western Greenland below 81° north latitude, and eastern Greenland below about 70° north latitude, are void of murids. This has been supported by the report that blue fox populations do not fluctuate as much as "lemming fox" populations (Banfield, 1974).

An examination of Hudson's Bay Company fur statistics for the different color phases of the arctic fox from 1848 to 1909 (figs. 16 and 17) may suggest some support for Braestrup's theory, since the white phase reveals an emergent spectral peak in the vicinity of 4 to 4.2 years ($f = 0.23$ to 0.25); whereas no single peak begins to dominate in the blue fox spectrum. However, the evidence may be considered by many to be somewhat equivocal, since the peaks are not as dramatic as that for, say, the Mackenzie lynx data. Since these data reflect fur returns for all of northern North America, with the exception of the Yukon, parts of British Columbia, northern Quebec, and the outer zones of the Mackenzie River basin (Elton and Nicholson, 1942b), and since most population fluctuations represent local fluctuations, varying differences in the timing of peaks between regions can seriously distort the picture for the total area.

Fortunately, a more convincing data series is available from the Moravian Mission fur collections from the coast of Labrador for 1834–1925 (fig. 18). The spectral peak emerging in the vicinity of 4 years ($f = 0.25$) reinforces the oscillatory autocorrelation function and the regular periodicity in the vicinity of 4 years seems beyond question. The

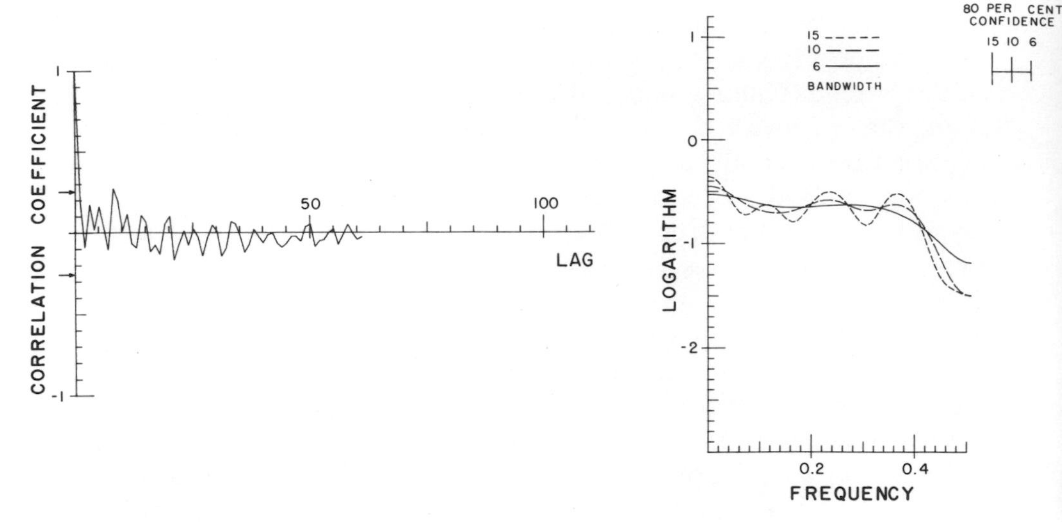

Figure 16. Autocorrelogram and autospectrum for *Alopex lagopus*, blue phase, 1848–1909. No single peak dominates the spectrum for these data, which represent returns for most of northern North America.

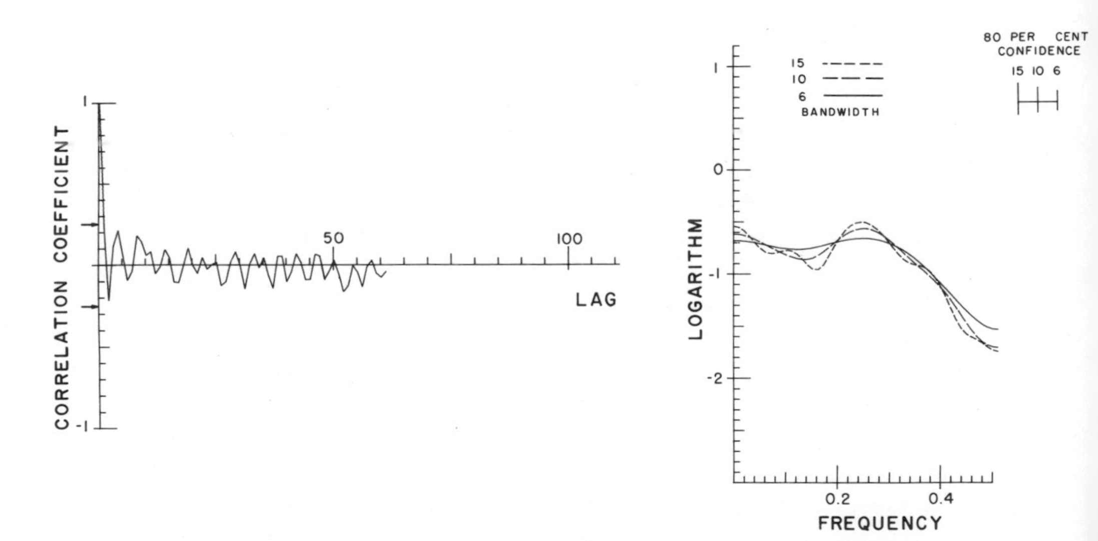

Figure 17. Autocorrelogram and autospectrum for *Alopex lagopus*, white phase, representing collections for most of northern North America, 1848–1909. An emergent spectral peak is suggested for a period between 4 and 4.2 years. Although this may support Braestrup's (1941) contention that the white phase of the arctic fox is more lemming-oriented than the blue phase (based on observations in Greenland), the evidence is equivocal.

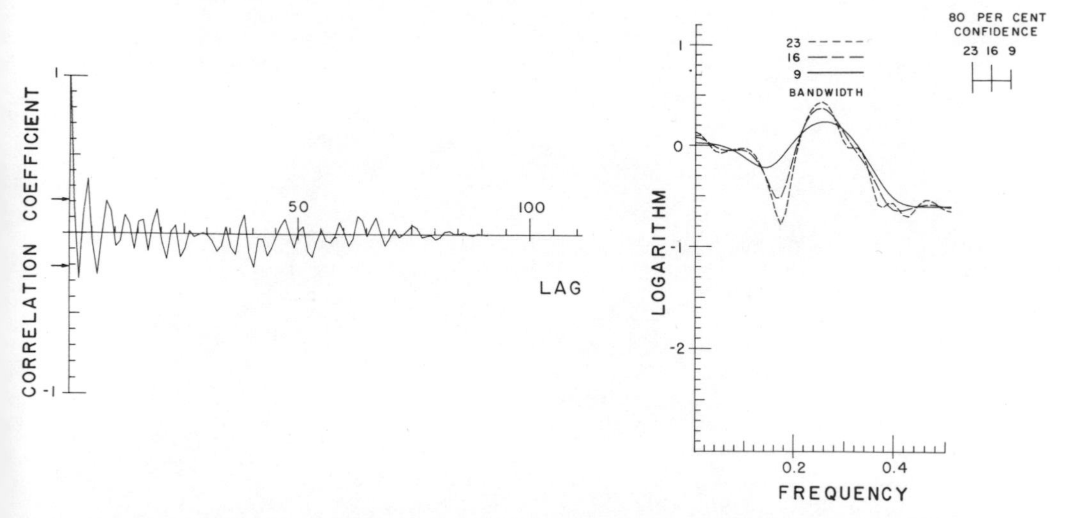

Figure 18. Fur returns for arctic fox from the Moravian Missions, Labrador, 1834–1925, offer an autocorrelogram and spectrum supporting strongly the contention that this species has regular oscillations, possibly due to the presence of a cyclic prey species.

Figure 19. Distribution of the arctic fox in Labrador and Ungava. (Based on Banfield, 1974.)

observation that the distribution of *Alopex lagopus* in this region (fig. 19) essentially overlaps that of its main prey item *Dicrostonyx hudsonius* (fig. 11) lends support to the idea that predator and prey populations are somehow inexorably connected in cyclic phenomena. Further support for regular cycles in arctic fox will be included in the following section on the red or colored fox.

Vulpes

The colored or red fox, *Vulpes vulpes* (Linnaeus) in northwestern Europe

Colored or red fox, *Vulpes fulva*

(fig. 20) and *Vulpes fulva* (Desmarest) in Canada (fig. 21), breed throughout the boreal forest zone, sometimes reaching into the tundra (Butler, 1945). The subspecies that inhabits the Scandinavian peninsula, *V. vulpes vulpes*, is distinguished by teeth that are larger and more robust than those in central and southern races, and by a skull size that is maximal for European foxes (Miller, 1912).

The prey of the colored fox depends on its particular habitat. In the far north the chief foods are lemmings and mice. Farther south, in the boreal forest, lemmings are entirely replaced by mice, and the snowshoe hare becomes the primary prey (Butler, 1951). Along the Alaskan coast fish and birds can be important food resources (Rue, 1969). When its primary food supply declines in abundance, the colored fox can switch to secondary prey, whether the primary prey be rabbits, rodents, or vertebrates in general (MacLulich, 1937; MacPherson, 1969). Sometimes, large amounts of vegetable food will be included in the diet (Burton, 1962), and insects are occasionally consumed (Rue, 1969). Extra food is often cached

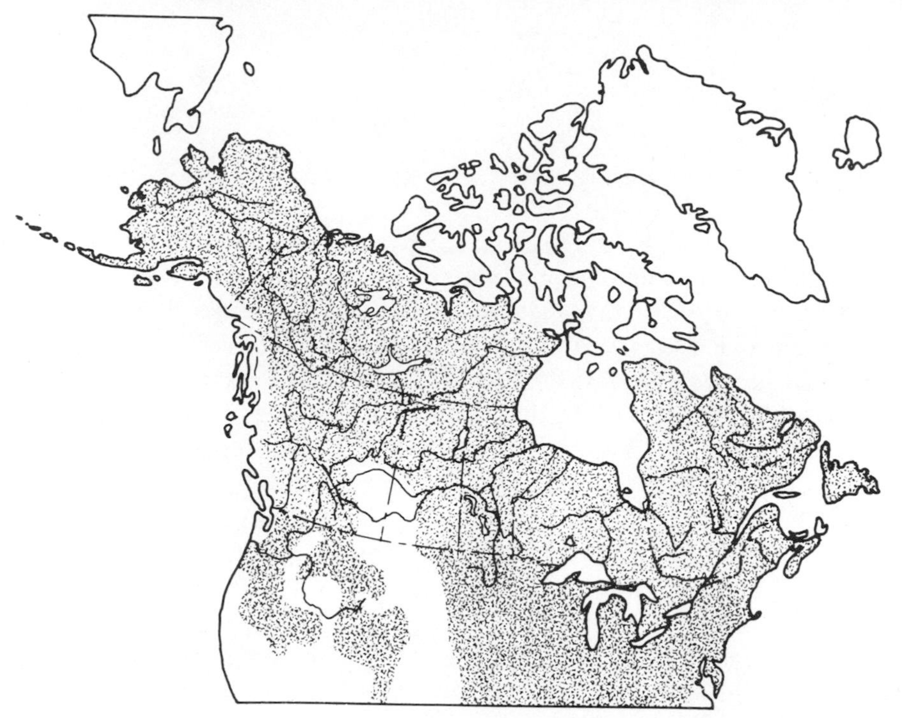

Figure 20. Distribution of the red fox, *Vulpes fulva*, in northern North America. (After Hall and Kelson, 1959.)

Figure 21. The colored fox, *Vulpes vulpes*, is ubiquitous in Scandinavia. (After van den Brink, 1967.)

(Rue, 1969). Unlike the arctic fox, colored foxes will not go onto the frozen sea in search of food (Elton, 1942).

The colored fox also differs from the arctic fox in that female maturity is not delayed. Breeding occurs mainly in January and February with delivery in April or May; the average litter is four pups (Burton, 1962). The vixen is fed by the dog-fox until the pups are born, and this may continue for several weeks thereafter. Weaning occurs at the age of 2 months (Rue, 1969), and the pups leave home in August (Burton, 1962). *V. fulva* can live up to 12 years, and the longevity of *V. vulpes* is said to be the same (Burton, 1962). *V. fulva* appears in three color phases, red, cross, and silver, the genetics of which have been the cause for much study, since they may be indicative of population movements (Butler, 1945, 1947, 1951, 1953; to be discussed).

On the Labrador coast, where the chief food of the red fox is lemmings and other microtines, clear evidence of a 4-year periodicity is evident in the fur catches of the Moravian Mission from 1834 to 1925 (fig. 22). Autocorrelations oscillate with a 4-year regularity and both positive and negative correlations are significant to at least 20 lags, and probably much further; and the spectrum shows a clear peak at $f = 0.25$, representing a period of 4 years. The fact that this predator reveals a 10-year periodicity

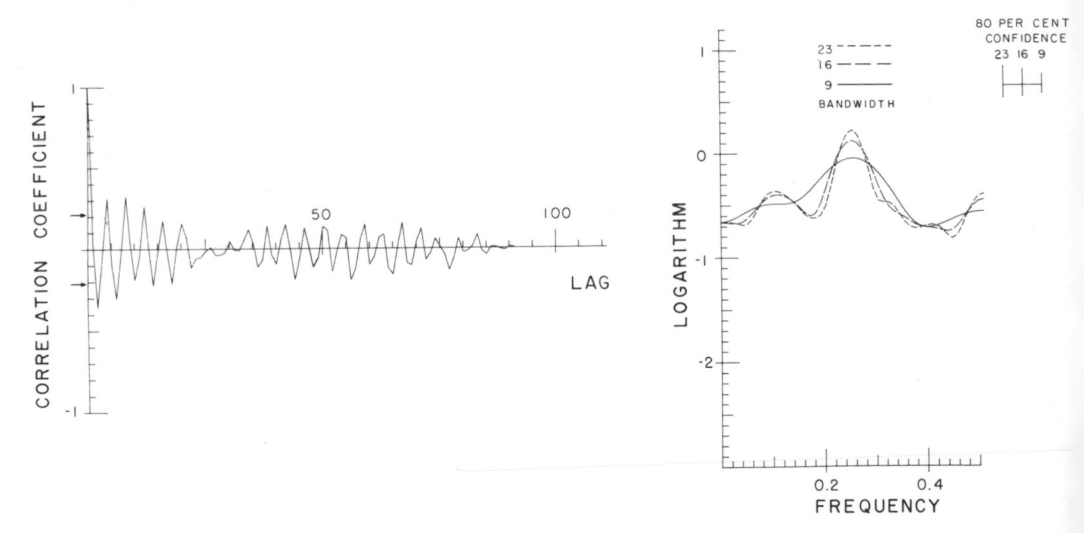

Figure 22. Serial correlogram and autospectrum for colored fox collections from Labrador Moravian Missions, 1834–1925. The spectral peak near $f = 0.25$ gives clear evidence of a 4-year periodicity for this species in this region. Ten-year periods in other parts of Canada (see fig. 35) suggest that this species may be tracking local prey fluctuations.

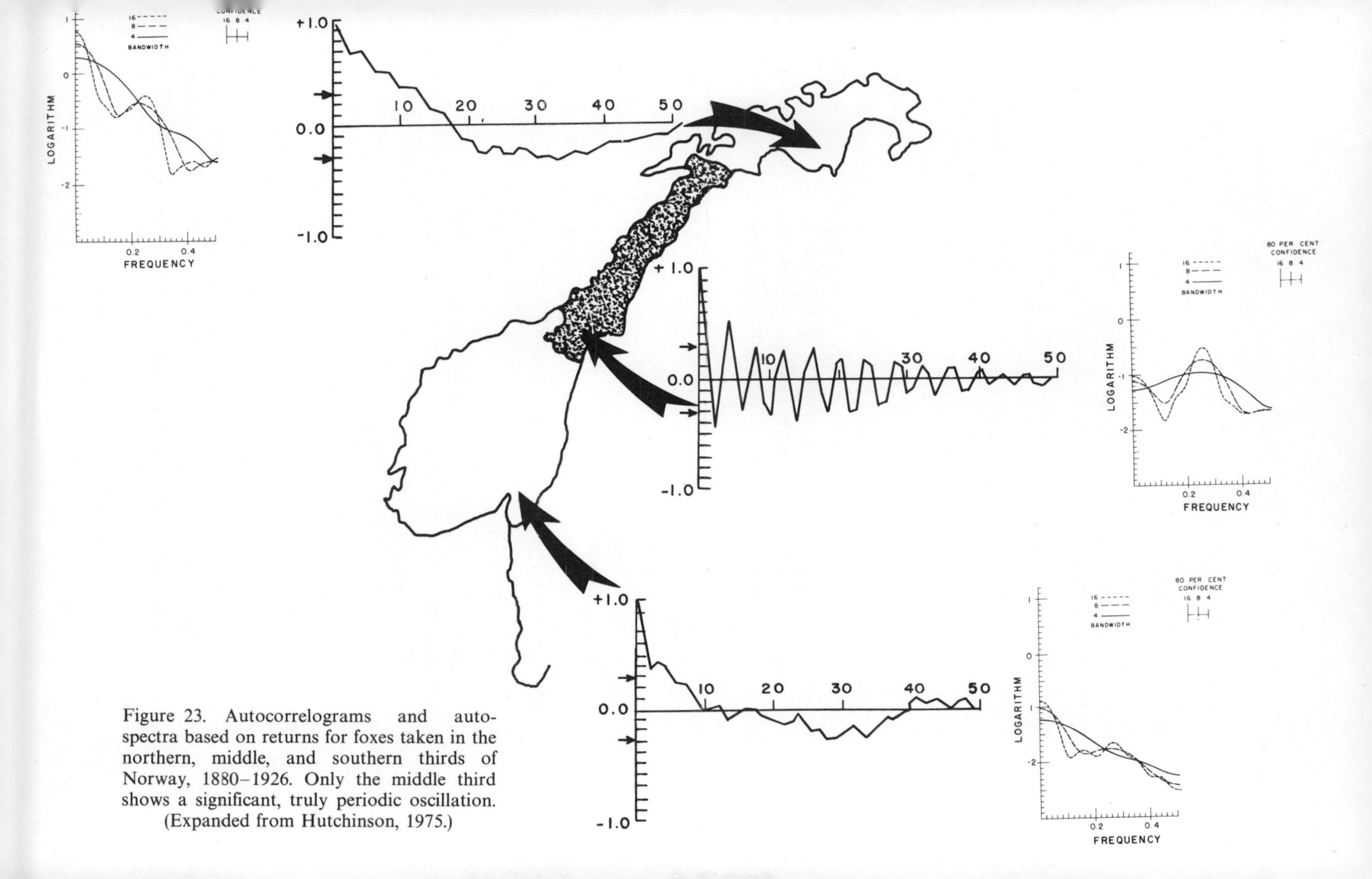

Figure 23. Autocorrelograms and auto-spectra based on returns for foxes taken in the northern, middle, and southern thirds of Norway, 1880–1926. Only the middle third shows a significant, truly periodic oscillation. (Expanded from Hutchinson, 1975.)

in the boreal forest may imply that predator cycles are in some way tracking prey cycles.

Bounty records in Norway were tabulated by Elton (1942) based on data organized by Johnsen (1929). These data represent both *Alopex lagopus* and *Vulpes vulpes* and were grouped by Elton into three regions, determined by information on lemming movements discussed by Collett (1895). Time-series analysis of these data are shown in fig. 23. Sizable population shifts in the northern and southern regions distort the autocorrelation functions and the spectra for these areas, while suggesting peaks near $f = 0.26$, are somewhat dominated by low-frequency power. No attempt was made to compensate for this difficulty by smoothing the data. The central region, corresponding to the area of Norway where lemming migrations most frequently occur (Collett, 1895), shows clear oscillations in the original data and in the autocorrelation function, and a clear spectral peak corresponding to a period of 4 years. A more detailed analysis of these data will be discussed when the question of the role of dispersal in population fluctuations is addressed, but clearly the existence of a regular periodicity, probably connected to the lemming flux, cannot be denied.

Mustela erminea

Mustela erminea, the ermine, is widely distributed throughout all of the northern part of North America (fig. 24) and Europe. Although their preferred home seems to be the boreal coniferous forest or mixed forests, they also inhabit tundra, meadow boundaries, shrubby river banks, and lakeshores. They occur from sea level to alpine tundra at an elevation of 10,000 feet (Banfield, 1974).

Female ermine mate in the early summer, but implantation is delayed until the next March, and the litter, usually four to seven in number, is delivered about 1 month later. The bulk of the ermine's food is voles, mice, and shrews, although rabbits, birds, amphibians, and some invertebrates are also eaten (Petersen, 1966; Banfield, 1974). On the tundra they also consume lemmings (Elton, 1942), and they can pursue these right into their underground passages; in tundra areas the lemming may be the ermine's only good food source in the winter (Elton, 1942).

A study of the Hudson's Bay returns for 1848–1909 (Jones, 1914) does not show any regular periodicity for ermines (fig. 25). This could reflect one of two things: (1) the data, which include all of Canada, may be dominated by ermines from forest areas where their major prey (voles or mice) are not cyclic; and (2) the ermine's ability to go under the snow for

Ermine, *Mustela erminea*

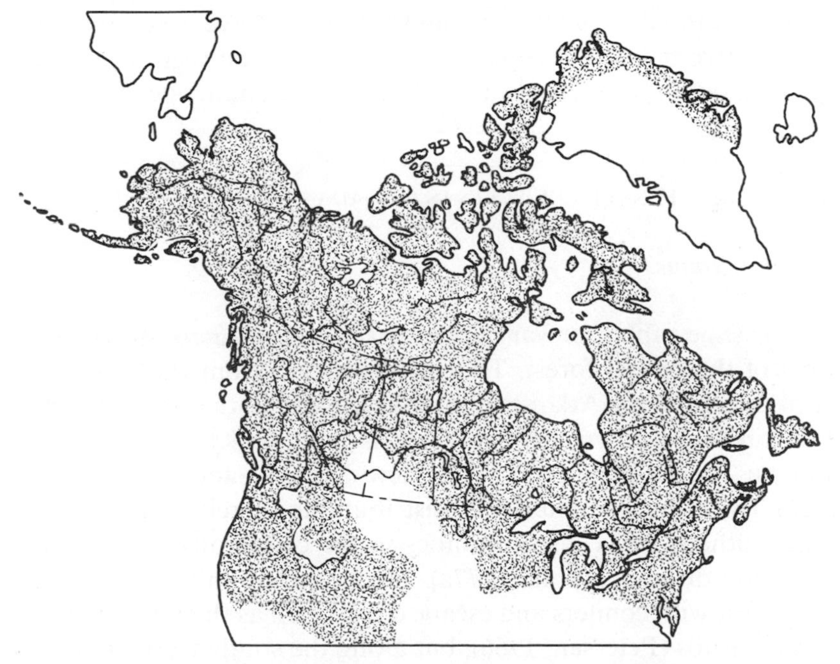

Figure 24. Distribution of ermine, *Mustela erminea*, in Canada. (Simplified from Hall and Kelson, 1959.)

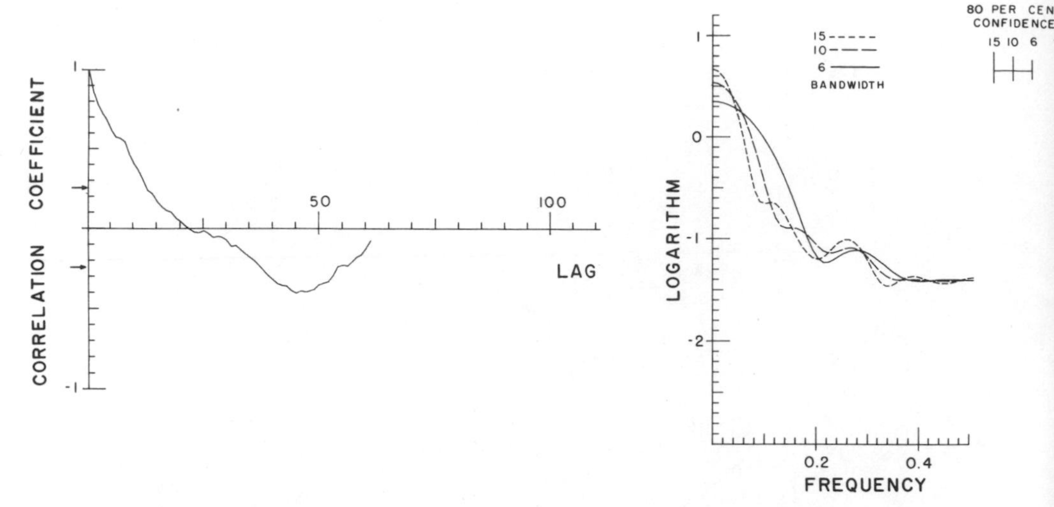

Figure 25. Serial correlations and smoothed autospectrum for ermine collections in Canada between 1848 and 1909 do not suggest a regular periodicity for this species.

lemmings, or other prey, means that its food supply is not limited to those seasons when lemmings appear frequently (this type of steady predation is suggested in Pitelka, 1973). It would be useful to have long-term data on this species from different regions (e.g., a purely forest region and a purely tundra region) to see if a cycle might become evident in one of these areas.

Evidence for Ten-Year Cycles in Canada

Lepus americanus

The snowshoe rabbit, or varying hare, *Lepus americanus*, is primarily an animal of the boreal forest. The species ranges from the west coast of central Alaska to the Atlantic Ocean, extending to the northern limits of the boreal forest down through the Cascade, Rocky, and Appalachian mountain ranges into California, New Mexico, and North Carolina (fig. 26). The central and northern distributions are relatively continuous, but in southern areas discontinuities in the coniferous forest make the distribution disjunct (Wolff, 1977a). The preferred habitat of the hare is mixed forest with conifers and escape cover, such as thick brush deadfalls and low shrubs (Petersen, 1966), but along the southern periphery of its range it is usually found in cedar or spruce swamps. Snowshoe rabbits are active in the evening, seeking herbs, shrubs, and other vegetable matter for

Snowshoe hare, *Lepus americanus*

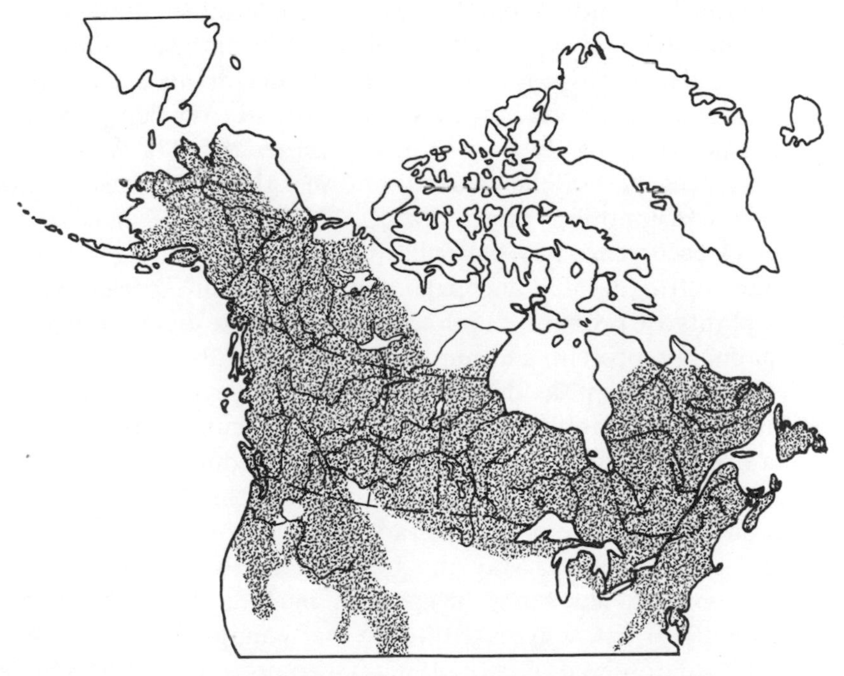

Figure 26. Distribution of the snowshoe hare, *Lepus americanus*, in Canada. (Based on Hall and Kelson, 1959.)

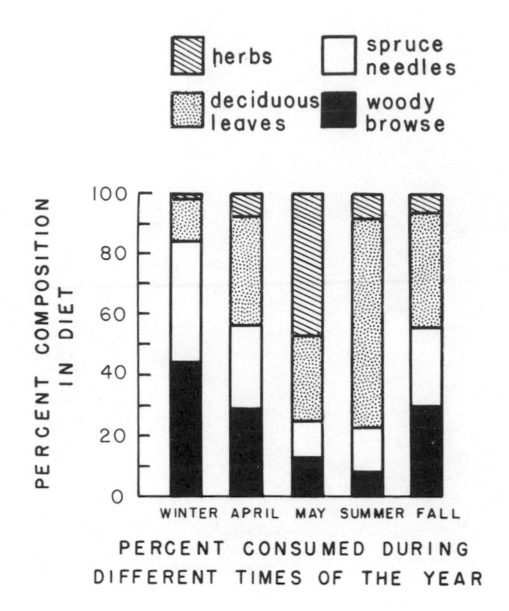

Figure 27. Food habits of the snowshoe hare. Bars indicate percent of woody browse, spruce needles, deciduous leaves, and herbs consumed at different times of the year. (Based on Wolff, 1978b.)

food. In an Alaskan study, Wolff (1977a, 1978b) determined that the hares' diet is seasonal, being composed primarily of herbs and deciduous leaves in the spring, deciduous leaves in the summer, deciduous leaves, spruce needles, and woody browse in the autumn, with a heavy dependence on spruce needles and woody browse in the winter (fig. 27). A survey of food preferences across the country showed that two basic factors determined consumption ratios (Wolff, 1978b): (1) the density and frequency of occurrence of a particular plant species in a given habitat, and (2) the nutritive value and palatability of a plant species. Several species of plants are probably required in the daily hare diet to provide the minimum energy, protein, and nutrients (Wolff, 1978b).

A shallow depression in the earth, snow, or vegetation suffices as a "nest" for snowshoe rabbits (Petersen, 1966). Does can produce as many as four litters a year. The majority of hares start breeding at the same time in any given locality and thus bear young at about the same time. Litters can be produced as early as April, although early litters do not occur in northern parts of Canada. May and June see the two major groups of litters, with only a few hares being born in August and September (MacLulich, 1957); litters average three to four young, although as many as 10 have been observed (Burton, 1962). The reproductive rate is higher during a population increase than during a decline (Banfield, 1974; Keith,

1974). Young are usually weaned 4 to 8 weeks after birth, but are not able to breed until the following spring (Anderson and Rand, 1945). The normal life span in the wild is probably 4 to 5 years (Banfield, 1974), although it may be significantly less (MacLulich, 1937).

The major predators of snowshoe hare are: great horned, great gray, and barred owls (*Bubo virginianus*, *Strix nebulosa*, and *Strix varia*, respectively), lynx, bobcat, red fox, coyote, wolf, mink, marten, and Indians (Banfield, 1974).

The best available data for snowshoe hare in Canada are from MacLulich (1937, 1957). The numbers refer to the quantities of pelts shipped by the Hudson's Bay Company to London for sale from 1849 to 1904, and are representative of the area around James Bay near Hudson Bay, since hare pelts were not valuable enough to ship from western Canada. A study of the logarithms of the data (fig. 28) reveals that distances between maxima range from 8 to 11 years, whereas minima are more regularly spaced every 10 years. The autocorrelation and spectral analyses suggest that the average cycle length is slightly greater than 10 years, and that the oscillations are relatively regular.

It is interesting to note that a similar analysis of game records for *L. americanus* collected by the Pennsylvania Game Commission (fig. 29) (from Keith, 1963, and Pennsylvania Game Commission, 1965; see Finerty, 1972) does not disclose any statistically significant periodicity. Andersen's (1957) study of the brown hare, *L. europaeus*, in Denmark similarly showed no indication of truly periodic variation. However, it is likely that the mountain hare, *L. timidus*, in the republic of Komi in the northeastern corner of European Russia, a taiga zone, does have long-term cycles (Naumov, 1972). These facts suggest that there may be something in the structure of certain boreal habitats that sets the stage for 10-year cycles.

For reference, a complete snowshoe hare cycle, as visualized by Keith (1974) and supported by Wolff (1977a), is summarized next. Approaching peak densities, an increasing interaction between the hare population and its overwinter (September–May) food supply of woody browse brings the forage availability to a critical level, where it is below that needed to support the population through the winter. Excessive browsing, sometimes including girdling, reduces subsequent annual increments of new growth either by killing the larger stems of woody plants or by lowering their vigor. This food shortage tends to persist for 1 or 2 years after the decline, since the decrease in hare numbers is offset by the decrease in new browse production and total standing-crop biomass.

Immediately preceding the peak winter of critical food shortage,

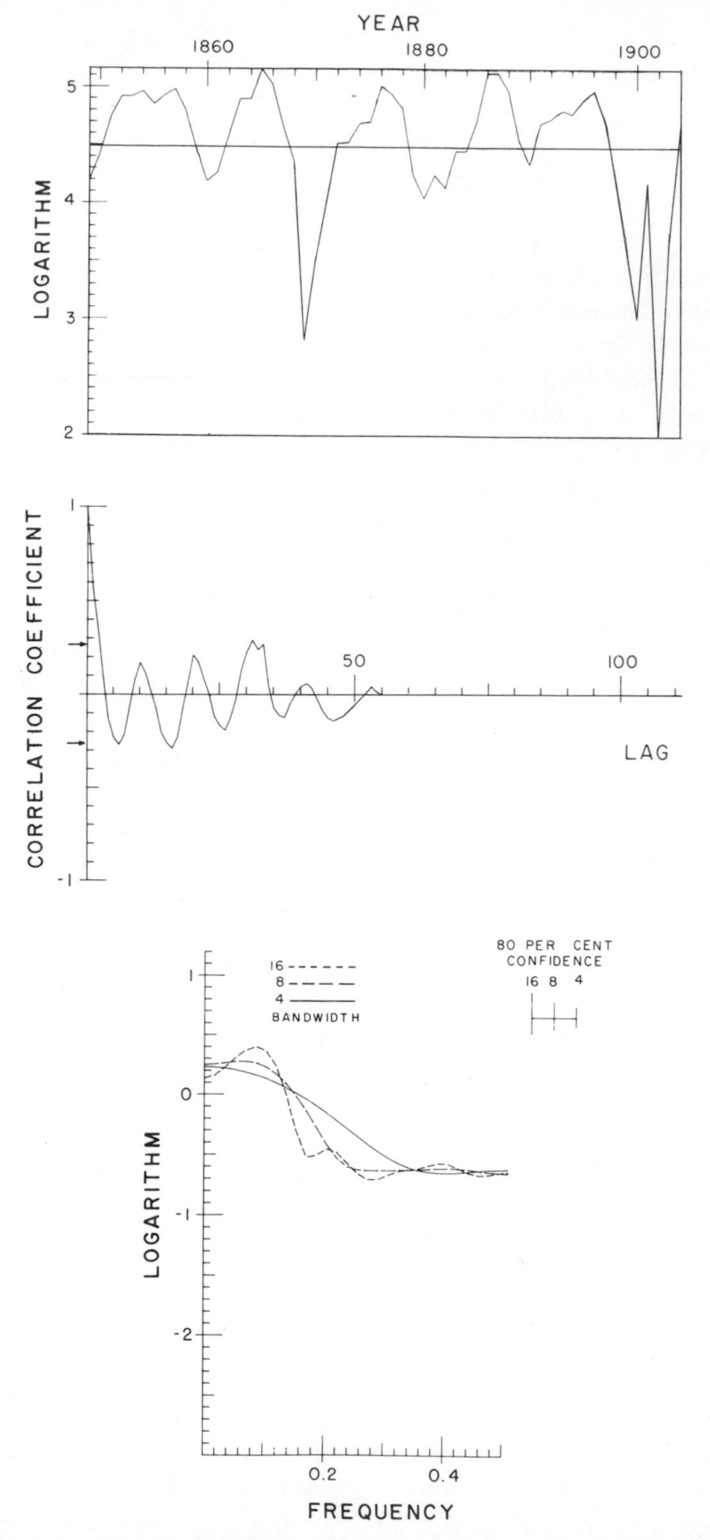

Figure 28. Time-series analysis for snowshoe hare in the James Bay area near Hudson Bay, 1849–1904. The original data suggest that zeniths are irregularly spaced, while nadirs are more regularly spaced. Autocorrelation and spectral functions suggest an average cycle length slightly greater than 10 years, and that oscillations are relatively regular.

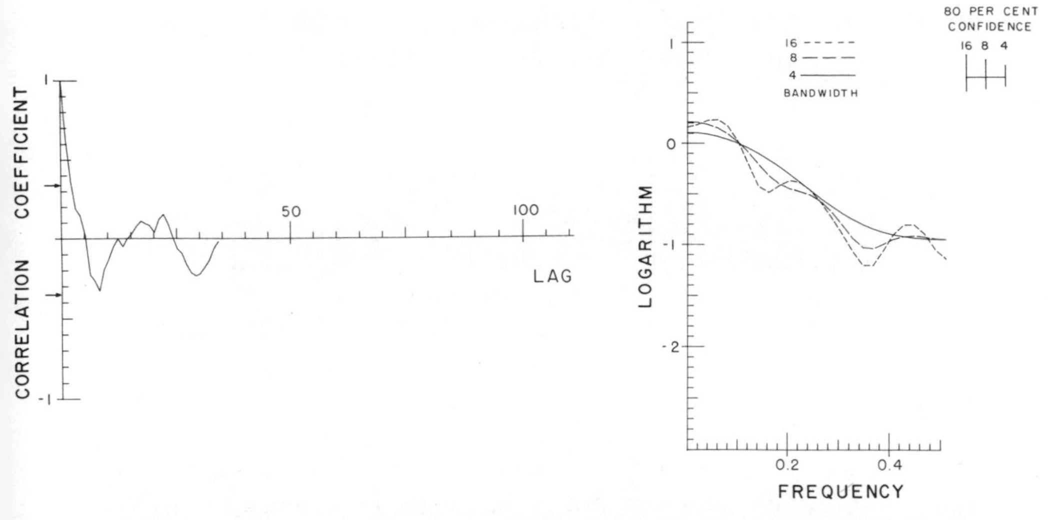

Figure 29. No statistically significant periodicity is suggested by this time-series analysis of game records for snowshoe hare collected by the Pennsylvania Game Commission between 1930 and 1964.

reduced availability of suitable browse and consequent decreased vigor of breeding adults results in lower reproductive rates and slower juvenile growth, which continue through the early years of the population decline while overwinter food shortage persists. Food shortage produces extremely high juvenile mortality in the peak winter.

Predator populations, having increased with the increased prey availability, still do not reach a level where they have a significant effect on peak hare densities. However, as hare populations decline, the predator–prey ratio increases to a point where predation becomes the dominant force in hare mortality. High predation depresses hare survival and population density below the potential dictated by the available food supply and therefore extends the period of decline.

The chief impact of weather is its effect on the hare–vegetation interaction, higher hare mortality being associated with lower winter temperatures and greater snowfall.

To this model Wolff (1976, personal communication) added: "During the winters of 1974–1975 and 1975–1976, the only hares remaining on my study areas were found in the densest habitats (refuges). It is my contention that dispersal of hares from these refuges into a mosaic of habitat types is important in determining the distribution and abundance of hares. The lack of buffer prey species in the northern environment forces obligate predators such as lynx, great-horned owls, and goshawks to fluctuate with the hare cycle." The similarity of this description to

Pitelka's (1973) description of lemming cycles at Barrow, Alaska, is clear, and suggests that a common model for lemming and hare cycles might be appropriate.

Lynx canadensis

"Of all the northern creatures none are more dependent on the Rabbits than is the Canada Lynx. It lives on Rabbits, follows the Rabbits, thinks Rabbits, tastes like Rabbits, increases with them, and on their failure dies of starvation in the unrabbited woods." Thus Seton (1911) colorfully described the extreme dependence of the Canada lynx on the snowshoe hare. This dependence has been verified in several studies, the most recent being in Alberta (Brand et al., 1976) showing that the percentage consumption of snowshoe hares when they were available approached 100 percent; when hares were not available, the lynx shifted to squirrels, ruffed grouse and other birds, and carrion, but these did not totally compensate for the low hare population, since a marked decrease in daily kill and consumption rates also occurred, setting the stage for starvation-related numerical responses (early postpartum mortality of kittens, reduced conception among yearlings, etc.). The lynx is, then, essentially single-prey-oriented and, although it will consume other small animals and birds if starving, it cannot exist successfully without the snowshoe hare upon which to prey (Elton and Nicholson, 1942b). Although usually sedentary, lynx can migrate long distances during food scarcity. It is interesting to note, also, that a lynx may go out of its way to kill a fox (Seton, 1911) but usually will not eat it (Rue, 1969).

Lynx inhabit heavily forested and swampy areas where snowshoe hares are available (fig. 30). They mate in January and February, and the young are born in March or April, the gestation period being about 60 days. A litter commonly contains four or five kittens (Moran, 1953b). Weaning occurs at about 2 months, but the kittens remain with the mother until the following spring (Burton, 1962), at which time they are capable of breeding.

Lynx can live as long as 11 years (Burton, 1962). Their major predators, besides man, are cougars and wolves (Banfield, 1974).

The fur returns for the Mackenzie River district, which occupies a large part of the northwest portion of the Northwest Territories (fig. 31; this is for 1897: changes in areas for the various Hudson's Bay districts were carefully documented by Elton and Nicholson, 1942b), have already been studied (figs. 3 and 6), and the periodicity is obvious. An examination of Canada-wide data for 1848–1909 (Hudson's Bay Company records from

Canadian lynx, *Lynx canadensis*

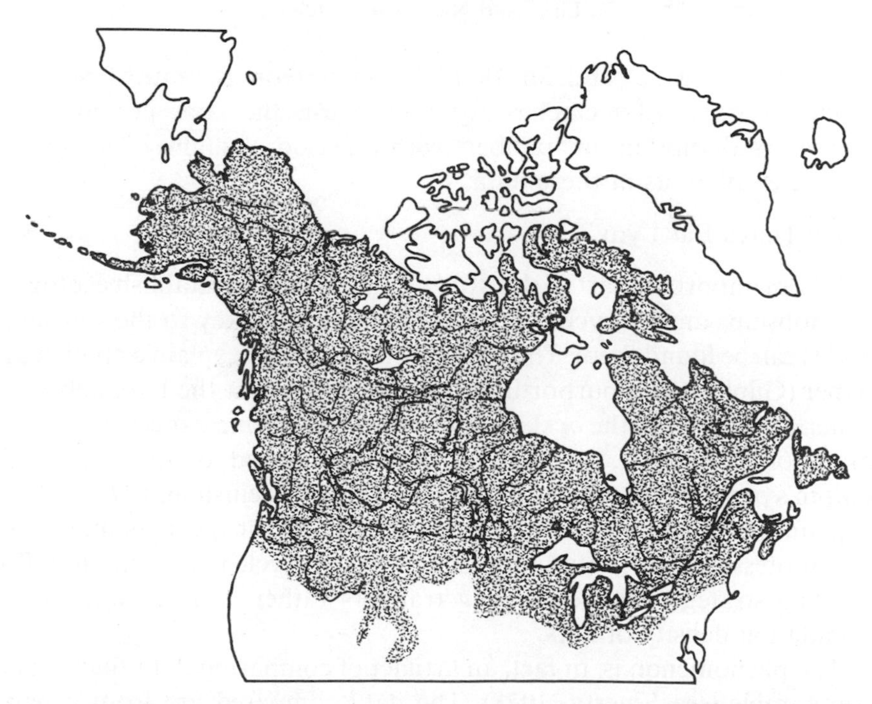

Figure 30. Distribution of *Lynx canadensis* in North America. (After Hall and Kelson, 1959.)

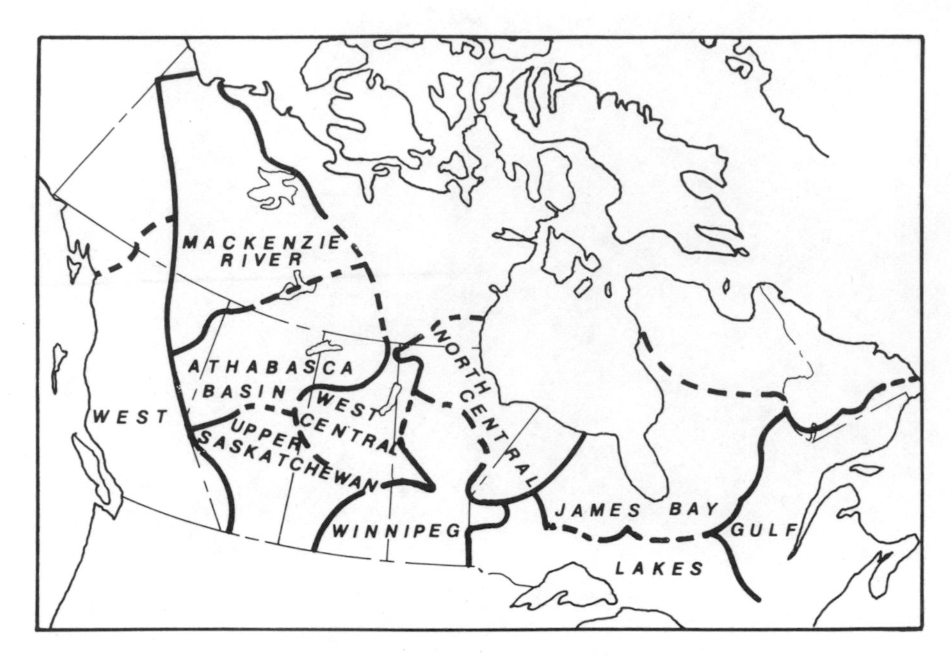

Figure 31. Organization of Hudson's Bay Company fur districts in 1897. (Based on Elton and Nicholson, 1942b.)

Jones, 1914; fig. 32) and for 1918–68 (Dominion Bureau of Statistics [1970] records of fur catches; fig. 33) suggests the same periodicity. A regular oscillation in fur numbers with a period of slightly less than 10 years clearly exists in these data.

Do Hares Eat Lynx?

It is an important part of this study to try to clearly establish ecological relationships among cyclic species to try to see if a key to the sources of cycles can be found there. Thus it seems necessary to explain a challenging paper (Gilpin, 1973) purporting to demonstrate that the lynx pelt cycle, instead of following the cycle in snowshoe hare as one would expect for a predator–prey cycle, is actually sometimes ahead of the hare cycle. Gilpin's paper, and a more recent commentary (Weinstein, 1977), tried to explain this phenomenon by suggesting that the fur trappers might be a major pressure on the population, and that the cycle in lynx might reflect hunting strategy changes among trappers rather than changes in the population density of lynx.

The phenomenon is, in fact, an artifact of comparing data that are not comparable (see Finerty, 1978). The data compared are from a paper describing the ecological role of Volterra's equations (Leigh, 1968). These

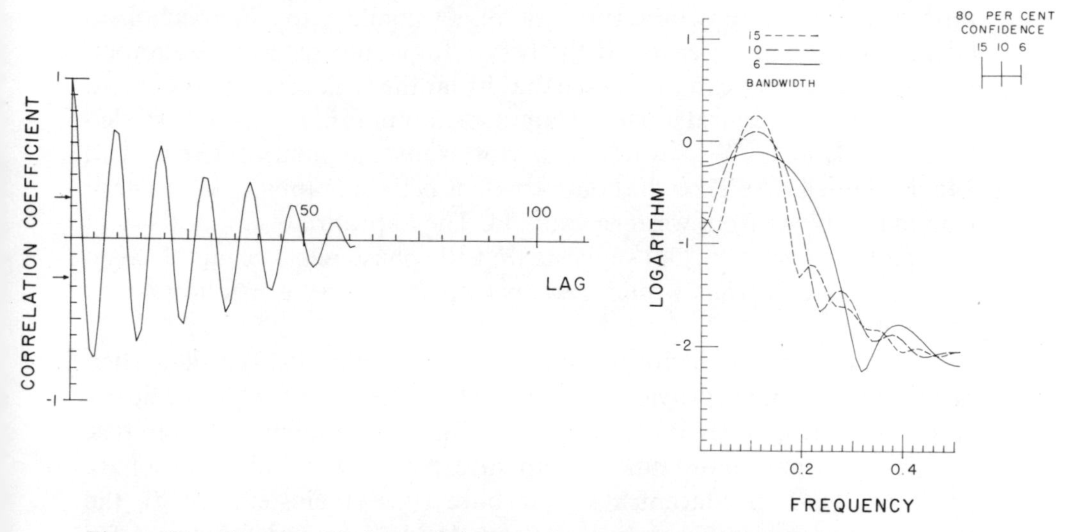

Figure 32. Canada-wide fur statistics for Canadian lynx, 1848–1909, demonstrate a clear periodicity.

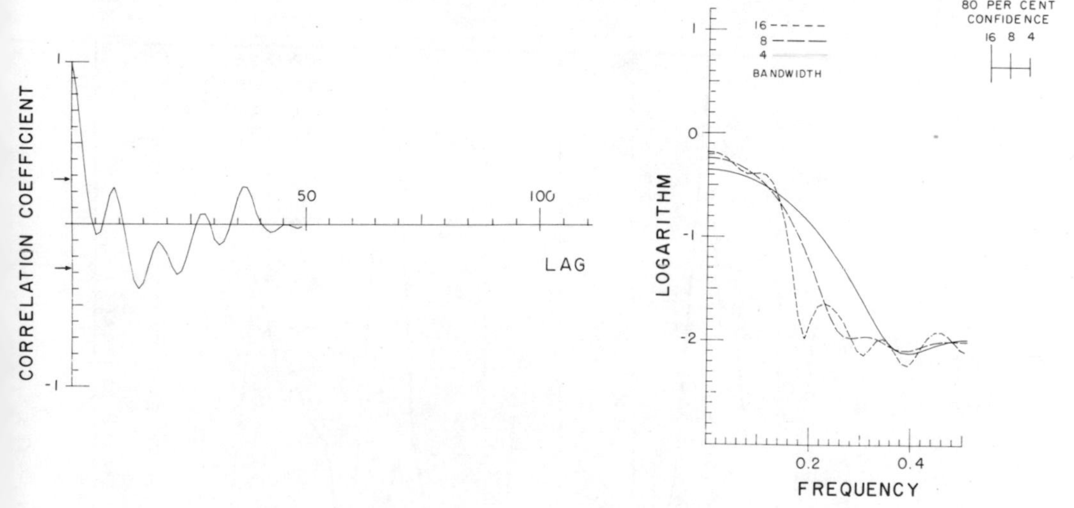

Figure 33. Canadian lynx data from the Dominion Bureau of Statistics, 1918–69, although variable, suggest that the lynx cycle still exists.

data were coarsely estimated, as Leigh clearly states, from a graph comparing hare and lynx fluctuations (MacLulich, 1937); the coarseness can be appreciated by referring to the original data (MacLulich, 1957, for snowshoe hare; Elton and Nicholson, 1942b, for lynx). The hare data are from the Hudson's Bay Company records for the area around James Bay, the southern end of Hudson Bay in eastern Canada. The lynx data represent Hudson's Bay fur returns for all of Canada. Elton and Nicholson (1942b) painstakingly separated the lynx returns into separate geographical regions, and one can clearly see that by far the bulk of the pelts derived from central and central western Canada; snowshoe hare was only traded from the Hudson Bay watershed, not from the interior (Mair and MacFarlane, 1908). Thus the comparison is between rabbits from eastern Canada and lynx from western Canada. The appropriate comparison of lynx and hare pelts (fig. 34) suggests that the phase relationship between lynx and hare is what would be expected for a predator-follows-prey situation.

Another important factor (Gilpin, 1973; Weinstein, 1977) is that when hares are abundant, providing easily obtainable food for the basically subsistence economies of many north Canadian Indian communities, trappers may have more time to trap valuable furs for trade; and if hare predators tend to concentrate near hare runs (Weinstein, 1977), the trapper may more efficiently trap both food (hare and lynx) and fur bearers. Years of low hare abundance might remove trappers from

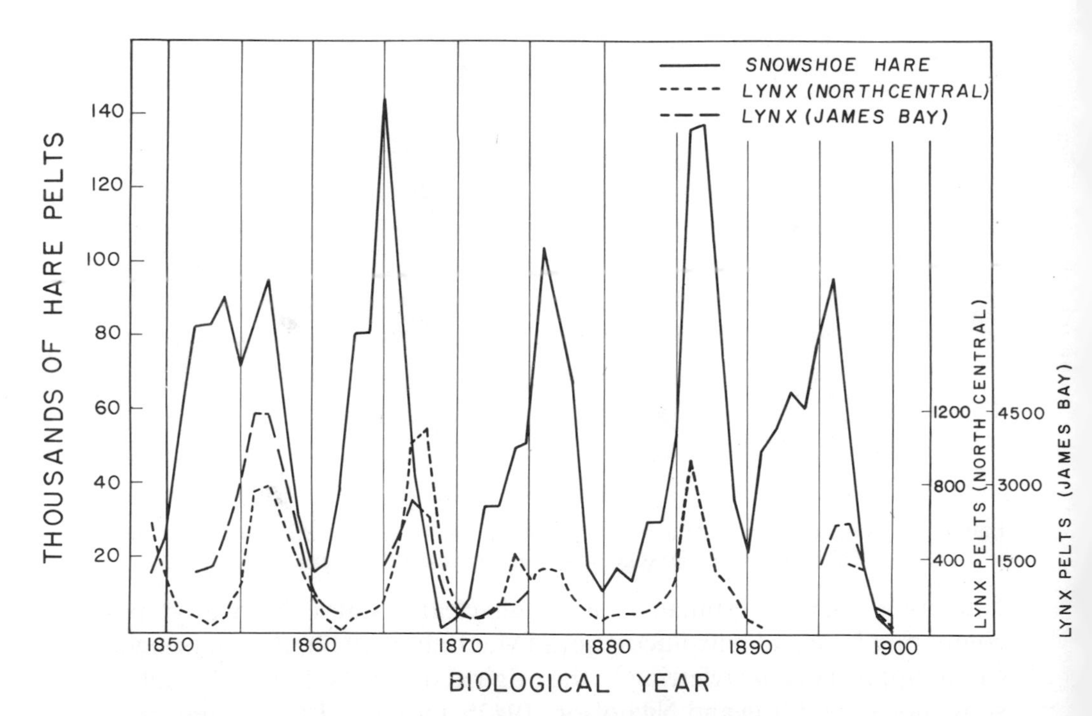

Figure 34. Comparison of snowshoe hare statistics with lynx data for the same approximate region (James Bay, North Central District) suggest the appropriate phase relationship for a predator-follows-prey situation.

habitats suitable for hare predators, causing the predator population figures (estimated from fur returns) to appear to drop more rapidly than is the actual fact. Although this might enhance population peaks and troughs, it would not generate regular periodicity. If the hare is cycling and the lynx pelt returns appear to cycle because the trappers are catching lynx incidentally in their search for food, it is possible that the lynx might not in themselves actually cycle (Weinstein, 1977), but the lynx have not been properly sampled for assessing this. Moreover, many naturalists have noted the abundance of lynx in peak years, and the magnitude of the lynx cycle seems too great to be explained solely by this trapping process (Elton and Nicholson, 1942b). The existence of regular periodicities in at least one other predator–prey pair, mink and muskrat (Elton and Nicholson, 1942a; Bulmer, 1974), plus cycles in several other mammal species (Finerty, 1972; Bulmer, 1974) probably could not be explained in a similar fashion.

Weinstein (1977) suggested that the biology of the lynx may not be able to explain the dramatic increases seen in the pelt records. It might be true that these occasional large rates of increase could not be explained by asserting that the lynx cycle is following hare abundance in a given area; however, the wide variation in phasing of peaks in the lynx cycles, suggesting an outward spread of lynx abundance from a central area (Elton and Nicholson, 1942b; Butler, 1953), implies that the lynx may be emigrating to follow the rabbit resources, and this could account for what is observed. Additionally, the average lynx litter size is two to three kittens, with a range as high as five kittens, with one litter per year (Banfield, 1974). Examining the distribution of the change in number of lynx pelts for the Mackenzie River district, Williamson (1972) observed that the average rate of population increase on the upswing is times two. Females mature in their first year and, allowing for a single annual breeding season, the rate of the population increase would be equal to parental survival plus the product of the proportion of females born and the rate of juvenile survival (E. L. Charnov, personal communication). If food is abundant, making parental and juvenile survival high, a female birth rate of slightly more than unity might be sufficient to double the population annually. Immigration from an epicenter (Butler, 1953; Watt, 1968) might account for yet higher rates, or discrepancies due to nonbreeding females. Thus the biology of the lynx could allow for the increases observed.

Although none of these possibilities is well enough documented to make definitive statements, it seems premature to abandon the possibility that lynx cycles may actually exist.

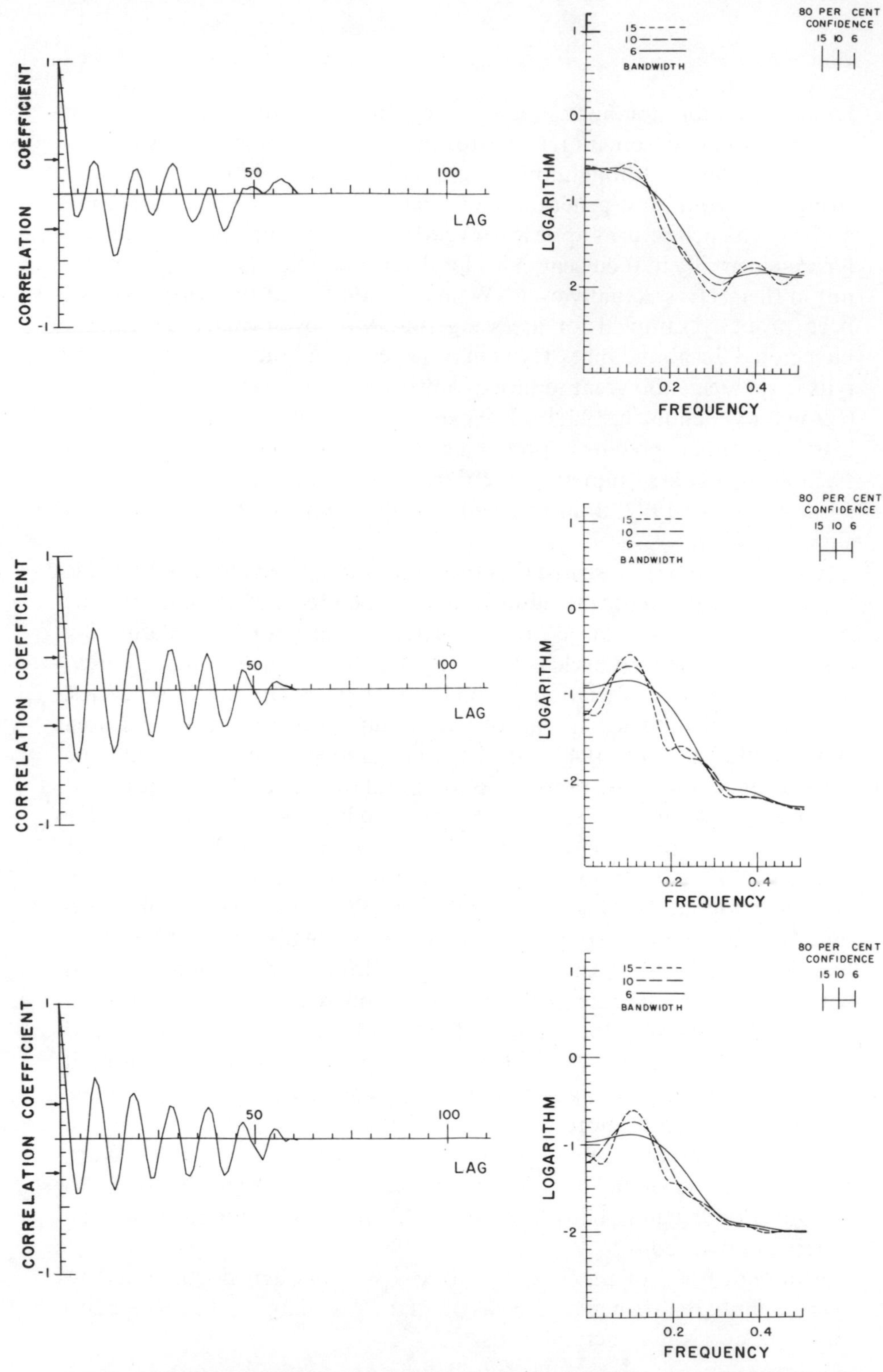

Figure 35. Hudson's Bay Company fur returns for the three color phases (red, top; cross, center; silver, bottom) of the colored fox, 1848–1909. These data, representing all of Canada, evidence a 10-year periodicity.

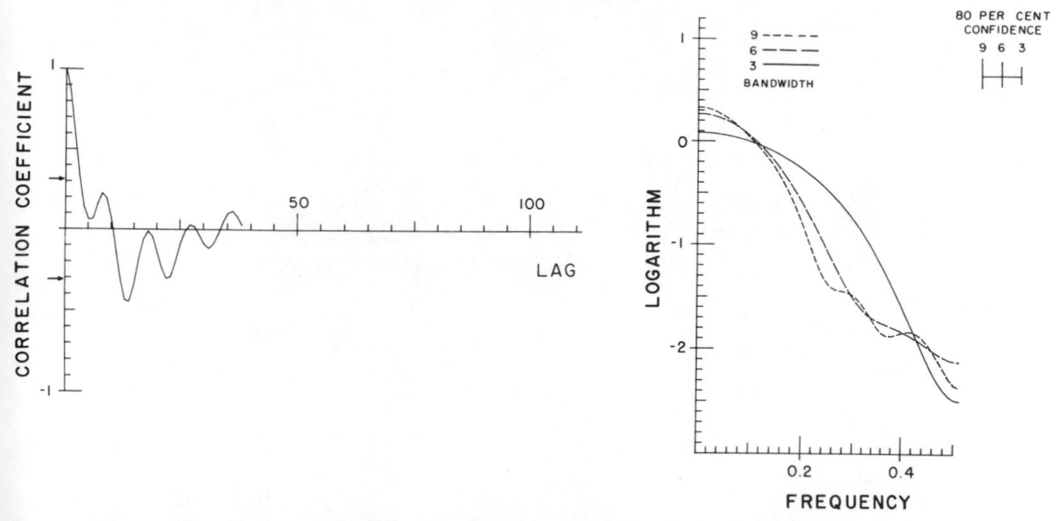

Figure 36. Data summed for the three Canadian prairie provinces of Alberta, Saskatchewan, and Manitoba, although not always directly comparable, still support the existence of a long-term cycle in red fox into the twentieth century.

Vulpes fulva

The biology of the colored fox and its three phases, red, cross, and silver, was discussed in the section on 4-year cycles. The Hudson's Bay Company returns for these phases for 1848–1909 (from Jones, 1914) clearly evidence a 10-year periodicity over all of Canada (fig. 35). Data tabulated by Keith (1963) for three Canadian prairie provinces (Alberta, Saskatchewan, and Manitoba), although they are not always directly comparable one to the other, support the existence of this long-term cycle in colored foxes into the twentieth century (fig. 36). These facts, combined with the obvious 4-year period in Labrador, is one of the strongest proofs that predator cycles are somehow determined by prey, although whether the relationship is one of interaction or of tracking remains to be clarified.

Martes americana

The pine marten, *Martes americana*, was said by Hewitt (1921) to be the most pronounced periodic furbearer in Canada after the lynx. Hewitt's examination of the nineteenth-century Hudson's Bay Company records (here from Jones, 1914) revealed a period for marten cycles in the neighborhood of 9.5 years, which is consistent with the time-series analysis (fig. 37). However, on the Labrador Coast, the Moravian Mission records suggest a short cycle (Elton, 1942). Why this disparity?

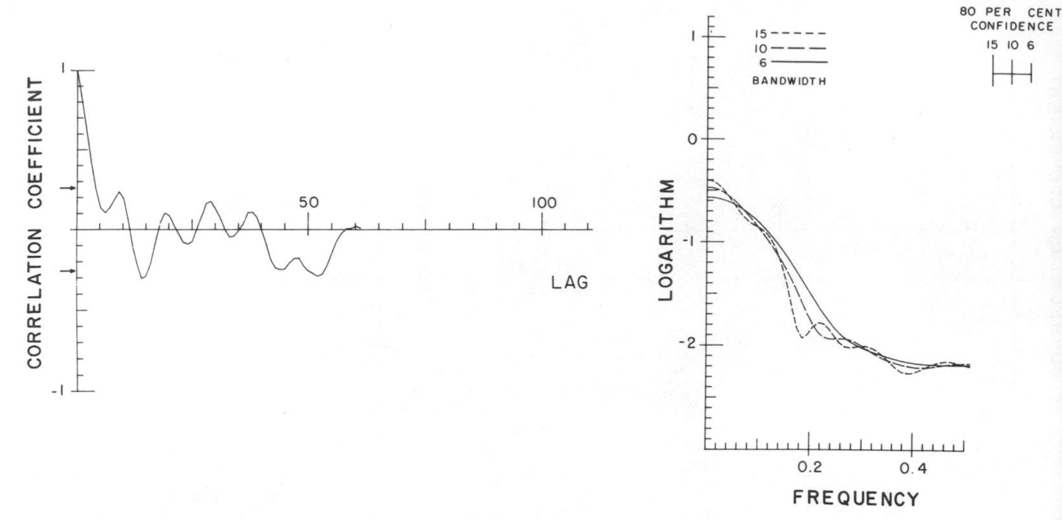

Figure 37. Hudson's Bay records for the pine marten, *Martes americana*, 1848–1909, in all of Canada, support Hewitt's (1921) suggestion that marten cycles with a period of about 9.5 years.

Martens are inhabitants of the boreal coniferous forest (fig. 38), and although they are considered to be tree dwellers, they spend a good deal of time on the ground hunting (Banfield, 1974). They are extremely agile in trees, where they pursue and catch red squirrels, but they are also efficient hunting on the ground, where they take mice, shrews, chipmunks, rabbits, birds and their eggs, amphibians, reptiles, insects, nuts, and fruit (Petersen, 1966). Microtines often form a large percentage of their diet, and when snowshoe rabbits are abundant, they may constitute a large portion of the marten's food consumption (Banfield, 1974). This broad diet probably determines the marten's population flux, ranging from 4-year periods where it feeds mainly on mice, to 10-year periods where hares are abundant.

Predation may also contribute to the periodic nature of the marten populations. Seton (1909) observed that when rabbits were scarce, martens (which are about the size of a mink) were preyed upon by lynx, fox, and fisher (which Seton felt explained the stories of their periodic "disappearance"), and coyotes, great horned owls, and golden eagles (*Aquila chrysaetos*) have also been named as predators, although none of these is thought to exert heavy predation pressure on the marten population.

Martens mate in July and August, and the females are polygamous. Owing to delayed implantation, the gestation period is 220 to 275 days, resulting in births the next March or April. The average litter is three or

Pine marten, *Martes americana*

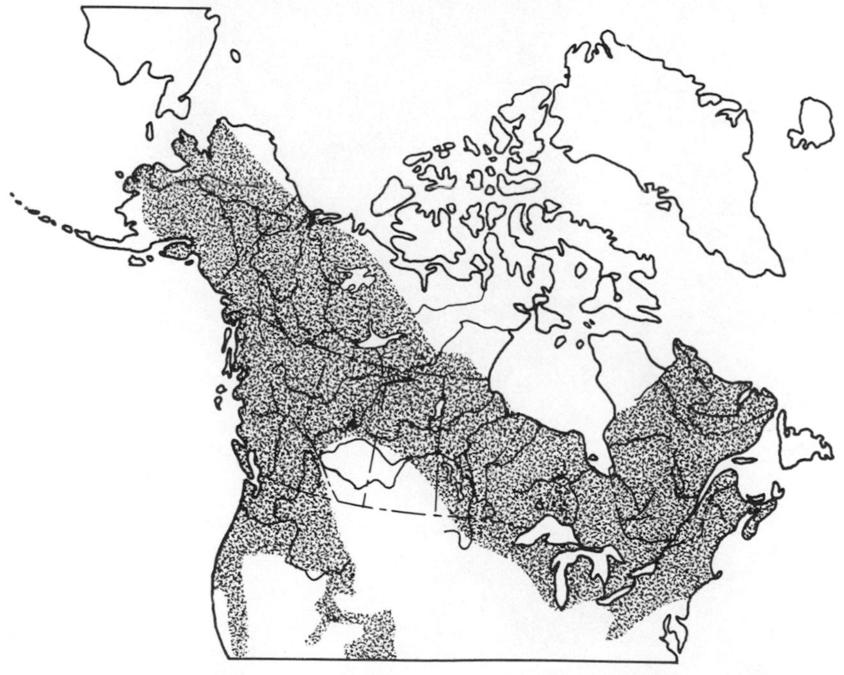

Figure 38. Canadian distribution of marten, *Martes americana*. (After Hall and Kelson, 1959.)

four (one to six per litter). While the females are sexually mature after 15 to 17 months, most do not bear their first litter until near their third birthday (Banfield, 1974). The average life span in the wild is not known, but probably is no more than 6 to 8 years (Petersen, 1966).

Martes pennanti

The fisher, *Martes pennanti*, has never been particularly abundant in any part of its range (Hewitt, 1921) and is now quite rare (Banfield, 1974).

Pennant's marten or fisher, *Martes pennanti*

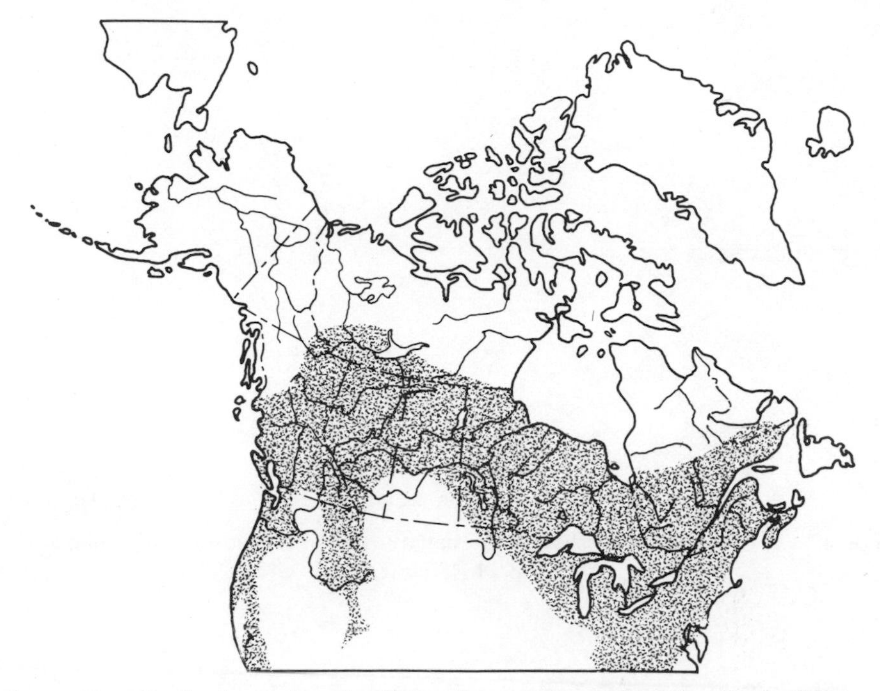

Figure 39. Distribution in Canada of fisher, *Martes pennanti*. (Based on Hall and Kelson, 1959.)

Fishers inhabit dense coniferous forests (fig. 39) preferably near a watercourse, although, unlike martens, they will venture into subclimax deciduous groves and old burns (Banfield, 1974). They travel a regular hunting circuit, ranging from 60 to 100 miles, which may encircle a home range 10 to 20 miles in diameter. Mammals constitute about 80 percent of the fisher's diet, and include sciurids, microtines, snowshoe hares (which Seton considered to be its favorite food; Seton, 1909), porcupines, and traces of mink and muskrats (Banfield, 1974). Food is often buried for later consumption. Carrion can also be an important dietary element.

The fisher becomes sexually mature at the age of 2 years and usually breeds in April. Delayed implantation lengthens the gestation period to about 1 year. The cubs, which may range in number from one to five per litter (average about three) are nursed for 7 weeks, and leave the nest after 3 months (Burton, 1962; Banfield, 1974).

A regular 10-year flux has long been attributed to fishers (consider Hewitt, 1921). This assertion is adequately supported by time-series studies of the Hudson's Bay Company statistics for the late nineteenth century (fig. 40), and data from 1919 to 1968 from the Dominion Bureau

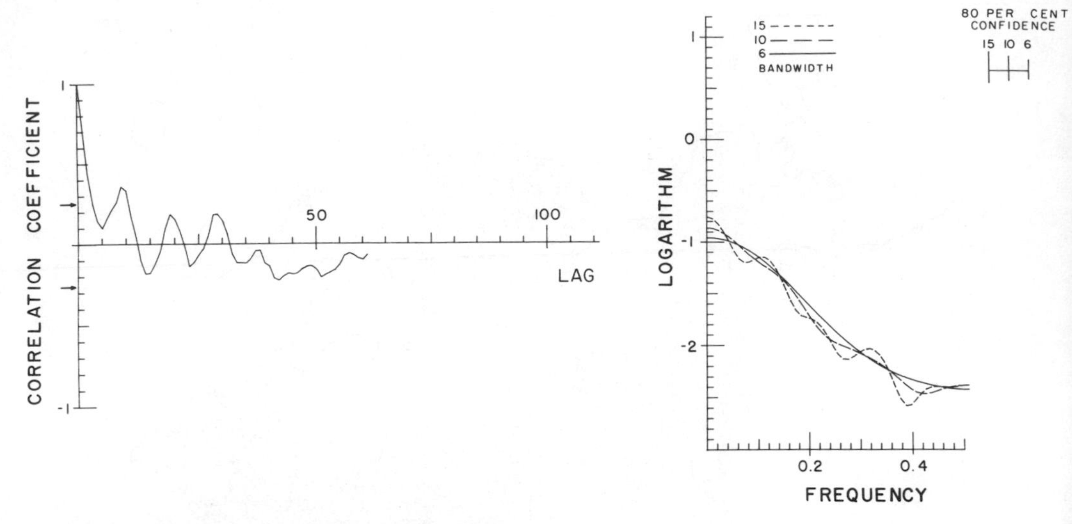

Figure 40. A 10-year cycle in fisher is suggested by analysis of Hudson's Bay Company records, 1848–1909.

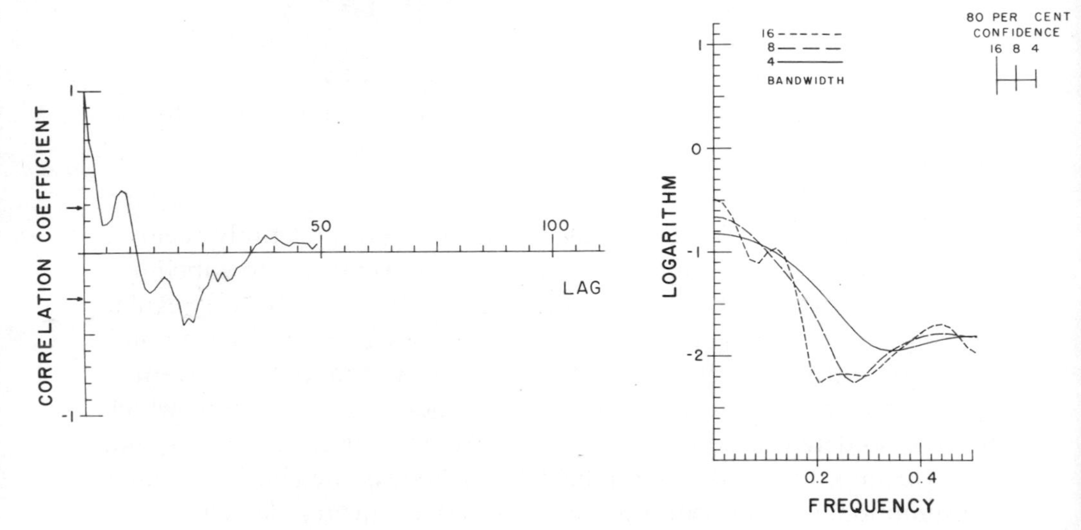

Figure 41. Dominion Bureau of Statistics records for fisher trapped between 1919 and 1968 weakly support the possibility for a continuing cycle in fisher populations.

of Statistics (1970) provides some weak support for the continuation of this cycle in the twentieth century (fig. 41). It seems reasonable to make the common assumption that the fisher is somehow connected with the cycle in the snowshoe hare.

Gulo luscus

The wolverine, *Gulo luscus*, formerly inhabited the boreal forests of North

Wolverine, *Gulo luscus*

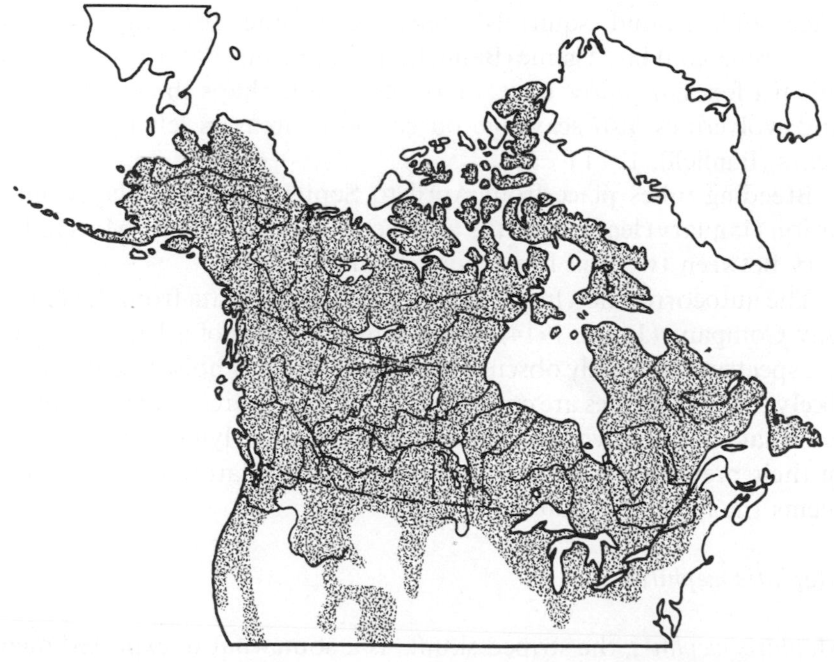

Figure 42. Wolverine distribution in Canada. (Based on Hall and Kelson, 1959.)

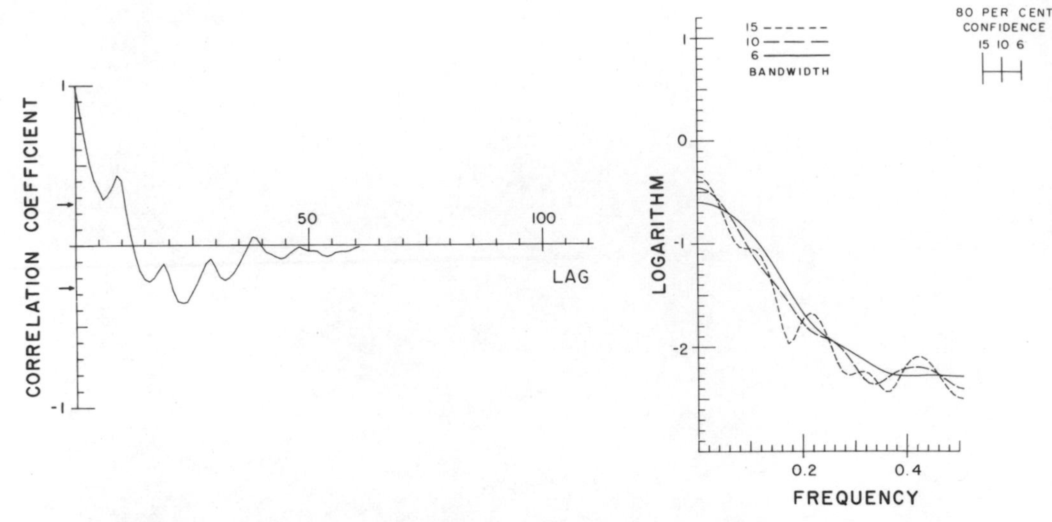

Figure 43. Although low-frequency noise obscures the spectrum, wolverine populations in Canada between 1848 and 1909 probably experienced 10-year cycles to some extent.

America,but is now found mostly on the Arctic tundra between the tree line and the coasts (Banfield, 1974) (fig. 42). It ranges widely (Seton, 1909), and is omnivorous, consuming edible roots and berries, small game (e.g., mice and ground squirrels), birds' eggs and fish, and sometimes porcupines and large game (Banfield, 1974). Seton (1909) also included on this list foxes, meadow mice, marmots, and perhaps the snowshoe hare, and wolverines also scavenge on caribou carcasses left by wolves and bears (Banfield, 1974.)

Breeding takes place from April to September, but delayed implantation (January) leads to births at the end of the following March. Litters vary between two and five young (Banfield, 1974).

The autocorrelation function for the wolverine data from the Hudson's Bay Company (Jones, 1914) shows some evidence of a 10-year cycle, but the spectrum is heavily obscured by low-frequency noise (fig. 43). It seems likely that wolverines are connected to cycles in hares, and perhaps foxes, and that their omnivorous diet obscures the underlying cycle in a portion of their prey species. Again, the concept of predator connected to prey seems the most reasonable here.

Mephitis mephitis

Mephitis mephitis, the striped skunk, is a somewhat unexpected member of this list of species, evidencing some cyclicity in the region of 10 years.

Skunks are distributed throughout the semiopen areas of mixed forests and open grasslands (fig. 44) and are basically omnivorous, feeding on fruits, insects, small mammals (field voles, deer mice, cottontail rabbits), grasses, carrion, corn, and nuts (Banfield, 1974; Petersen, 1966). They hibernate in winter, awakening to mate in late February–early March and to bear a litter of from two to ten kittens in the middle of May. None of these facts seems to explain why there is some evidence of a peak near 10

Skunk, *Mephitis mephitis*

Figure 44. Distribution of *Mephitis mephitis*, the skunk, in Canada. (After Hall and Kelson, 1959.)

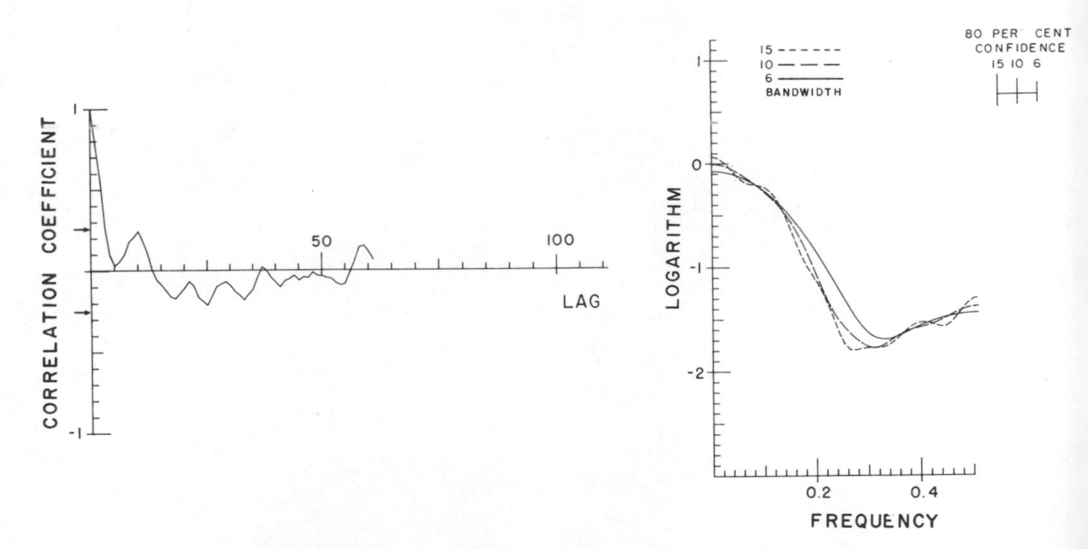

Figure 45. Weak evidence in the serial correlations, and weaker evidence in the spectrum, suggest that the skunk is somehow connected to 10-year oscillations in Canada. Based on Hudson's Bay Company data for 1848–1909.

years in the autocorrelogram (fig. 45) and the spectrum (despite the large amount of low-frequency noise).

The key, once more, seems to be predation, because the major predator of the skunk, the great horned owl (*Bubo virginianus*), is a major predator of snowshoe hares, and it is likely that the appearance of a 10-year peak is due to the predation pressure on the skunk when hares are at a low ebb.

Ondatra zibethicus

Phylogenetically, the muskrat, *Ondatra zibethicus*, is a member of the subfamily *Microtinae*, making it a close relative of the lemming. It is interesting to try to understand, therefore, why this species evidences cycles in the 10-year category while its near relatives are in the 4-year category.

Muskrats are widely distributed throughout North America (fig. 46) and are found in a wide range of aquatic environments: lakes, rivers, sloughs, ponds, and marshes. Depth of water is important to the muskrat because if the water is too shallow, it will freeze to the bottom in the winter and leave the muskrat without possibilities for travel or food, and if the water is too deep, the submerged vegetation necessary for winter survival will be missing (Banfield, 1974). Summer foods include bleached ends of reeds, lilies, stalks, and sedges, and clams, fish, insects, and freshwater mussels (Seton, 1909; Banfield, 1974).

Large population shifts occur in the autumn and spring. The autumn migration appears to be associated with abandoning dry ponds which will be useless the coming winter, but the spring migration is more pronounced, and appears to be associated with population dispersal driven by the need for excess females to find suitable dens and mates. These periods of migration leave the muskrat particularly susceptible to predation. Mink is probably the most serious predator, but foxes, coyotes, wolves, black bear, and lynx all enjoy muskrats. Several large avian raptors will kill muskrats if they are exposed in marshes, and pike and snapping turtles are a threat to young muskrats when they are swimming. But the mink can follow the muskrat into his den and is the major source of predation (Seton, 1909; Banfield, 1974). In a good habitat, however, none of these predators, nor heavy trapping, seems to affect population numbers seriously (Petersen, 1966).

Muskrats normally have two litters per year in northern Canada (Banfield, 1974) and may have three in eastern Canada (Petersen, 1966). In Louisiana, the southern part of the muskrats' range, litters may number three to six in a year but litters are small (as low as an average of 2.4),

Muskrat, *Ondatra zibethicus*

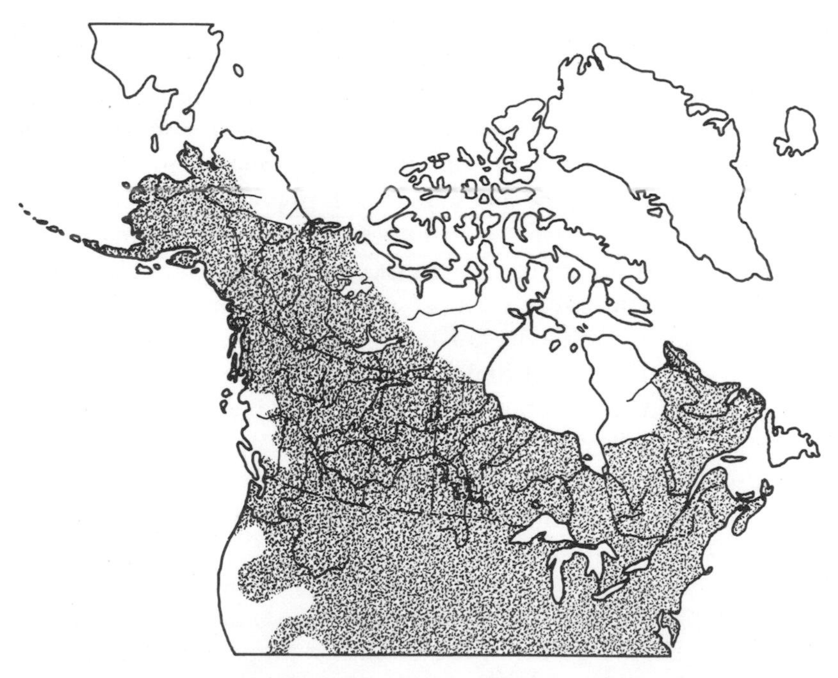

Figure 46. Muskrat distribution in Canada. (Based on Hall and Kelson, 1959.)

whereas in northern Canada they are larger (7.1). In Canada the breeding season is limited to March–September, whereas in the southern portions of its range, muskrat can breed throughout the year. In the north juveniles mature in about 1 year, whereas in the southern United States juveniles mature in about 6 months. The average life span of a muskrat is thought to be about 3 years (Banfield, 1974).

Data for muskrat collections from the Hudson's Bay Company were carefully sorted by Elton and Nicholson (1942a), who determined that trading practices in the first half of the nineteenth century made the returns for less expensive furs highly questionable as indices of population status. However, the last half of that century is acceptable and an analysis of the returns for all of Canada for 1848–1909 (from Jones, 1914) is shown in fig. 47. The autocorrelation function especially supports the observations (Seton, 1909; Hewitt, 1921) that muskrats are in fact cyclic.

Elton and Nicholson (1942a) clearly demonstrated that the fur returns of the Northern Department (Alberta, Saskatchewan, Manitoba, Mackenzie River basin, and part of western Ontario) accounted for most of the periodicity, although they made it clear that they did not wish to suggest that this implied an absence of cycles in some districts in other parts of Canada. They also observed that in Saskatchewan between 1915 and 1935, the muskrats in the north were abundant when those in the south were scarce, and vice versa. Butler (1962) pursued this observation

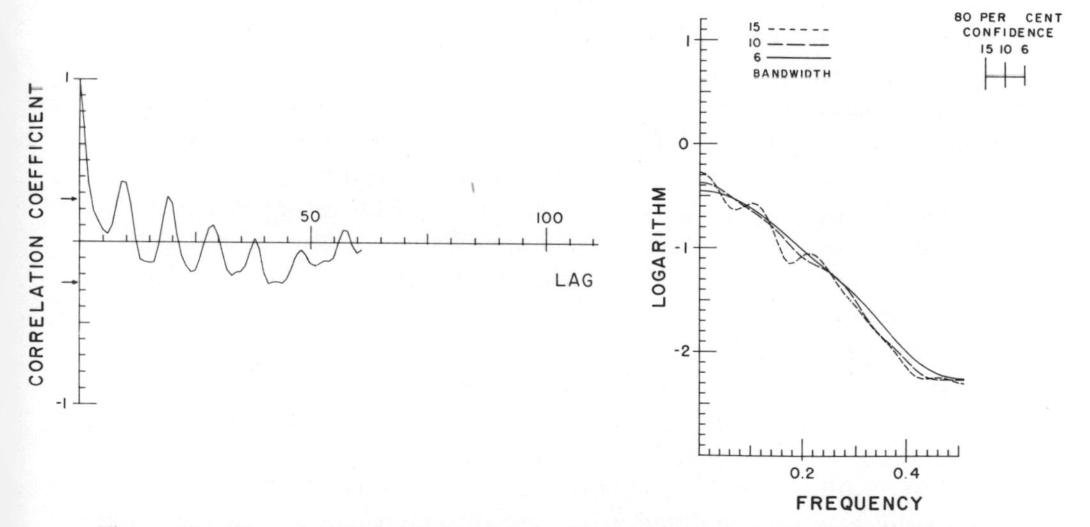

Figure 47. The autocorrelation function in this time-series analysis of muskrat populations in Canada, 1848–1909, especially supports the contention that muskrats evidence a 10-year cycle.

by tabulating fur returns from the Hudson's Bay Company and the Dominion Bureau of Statistics for 1915–60, and sorting these into three major regions for Saskatchewan: (1) the northern Canadian Shield country, a country of low relief and abundant lakes and rivers where the muskrat population comes mostly from small bays and shallow deltas, each of which rarely supports more than about 100 rats; (2) the central aspen grove region with fertile soil and few rock outcrops, an area usually watered twice annually by the flooding of the Saskatchewan River, and an area with extensive marshlands capable of supporting hundreds of thousands of muskrats; and (3) the southern prairie, where muskrats live in numerous sloughs and are greatly affected by rainfall, since drought will dry the sloughs.Analyzed in this manner, the muskrat fur data suggested interpeak intervals of 6 years in the prairie and aspen grove sectors, and average intervals of 10 years in the northern Shield. Highest catches in the prairie area showed correlation with two previous wet years, but precipitation did not correlate with either of the other areas in any significant manner.

How can these differences be explained? Butler thought that the most likely explanation is climatic. Differences in summer production and winter losses are two of the main reasons for variations in muskrat populations. Summer food quality, rather than quantity, is a major factor in summer production, and this is affected by nutrient influxes caused by a flood of silt-laden water or drying out and reflooding (allowing for accelerated plant decomposition during the dry period). Thus any major variation in the precipitation regime could knock a cycle out of phase, or even generate a different periodicity. Mark Boyce (1977, personal communication), after doing spectral analysis of muskrat returns for individual states and provinces in North America, observed that "states or provinces with more variable precipitation regimes generally have muskrat populations exhibiting poorer fits to a Fourier-series model of a particular harmonic." Boyce's preliminary analysis also suggested that there may be a wide range of period lengths depending on the region of the cyclic population; perhaps it will be possible to eventually connect this with the period of recovery of nutrients.

Most important, the muskrat in the northern forest evidences a period of population fluctuation of about 10 years. Whether this is because the muskrat is an alternate prey when snowshoe hares are unavailable (evidence will be presented later to suggest that muskrats and hares may not peak at the same time) or because of the structure of their habitat ("patchy island") and the period of nutrient recovery (which will be considered in a later section) may not be clearly determinable at our present level of knowledge.

Mink, *Mustela vison*

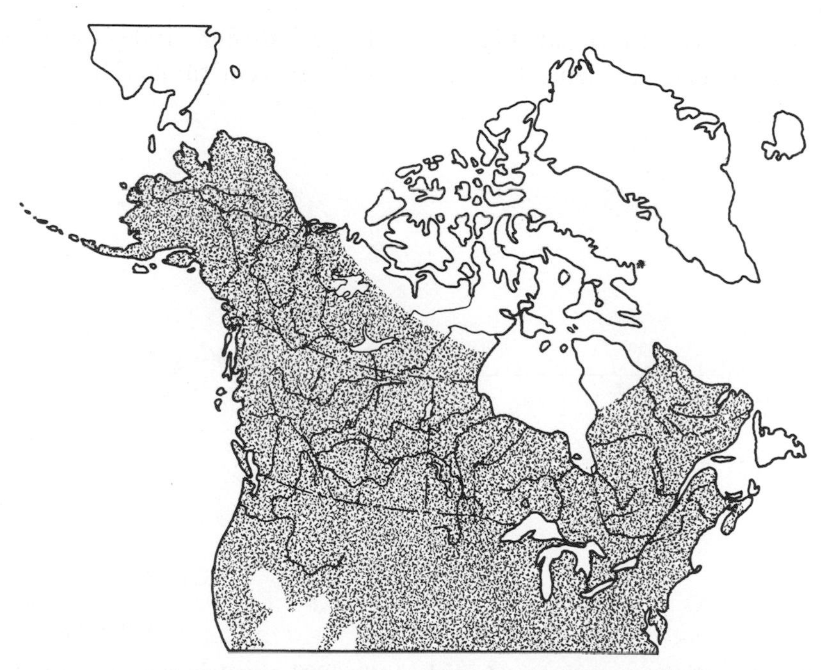

Figure 48. Mink distribution in Canada. (Based on Hall and Kelson, 1959.)

Mustela vison

The mink, *Mustela vison*, is widely distributed throughout Canada except in the northern part of the Ungava Peninsula (fig. 48). It is an inhabitant of the borderland between the water and the woods, living along streams, at the edges of ponds and lakes, and along the seacoast. The mink is essentially a weasel that has become adapted for an aquatic existence, and it is at home on the land as well as in the water.

Breeding begins in females at about 1 year, and in males at 18 months. The breeding season is from late February to early April, but because of variable delayed implantation births are confined to a period in late April and early May. Only one litter, with an average of five kits (range of one to six), is born per year.

The bulk of the mink's diet is composed of small mammals and fish. Among small mammals, the muskrat is probably the most significant prey (although rabbits are sometimes consumed; Seton, 1909). When juvenile muskrats are dispersing in the autumn they are especially vulnerable to mink predation. The life span of the mink is 3 to 5 years, on the average.

The mink's predators include the great horned owl, bobcat, red fox, coyote, wolf, and the black bear (Banfield, 1974; Seton, 1909; Petersen, 1966).

Hudson's Bay mink data (fig. 49) for the period 1848–1909 (Jones, 1914) show quite clear evidence for a cycle in the 10-year range (Hewitt, 1921,

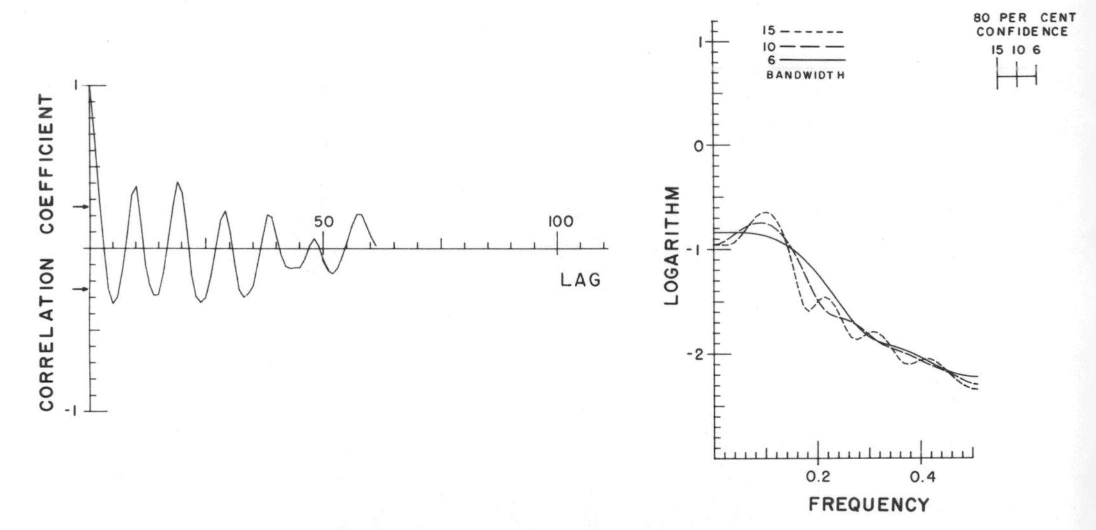

Figure 49. Mink populations in Canada, 1848–1909, reveal an extraordinary 10-year regularity in this time-series analysis.

suggested an average cycle length of 9.7 years). Examination of phase differences among species (to be discussed) suggests that it is more likely that the mink flux is determined by its relationship to the cycling muskrat than by its relationship as an alternate prey for snowshoe hare predators, although both of these are probably significant. Examination of twentieth-century fur returns for Saskatchewan and Manitoba (Keith, 1963) supports this contention.

Canis lupus and *Canis latrans*

The gray wolf, *Canis lupus*, and the coyote, *Canis latrans*, were combined by the Hudson's Bay Company under the heading "wolf." Mair and MacFarlane (1908) suggested that at least half of these furs were coyote in the first half of the period recorded by Jones (1914) (i.e., 1848–78), and Bulmer (1974) suggests that the proportion of coyote may have increased in the latter half (1879–1909) because of the marked decline in wolf numbers after the elimination of the plains bison (Hewitt, 1921). By dividing the series in two parts, Bulmer found a significant periodicity in the last half of the data, but no periodicity in the first half, which probably implies that wolves are not cycling, but coyotes are (fig. 50). Keith's (1963) graphs of coyote fur returns from three prairie provinces, which are the stronghold for the coyote in Canada (fig. 51), support the idea of a cycle in this species (fig. 52). Once again, this is probably due to preferred prey

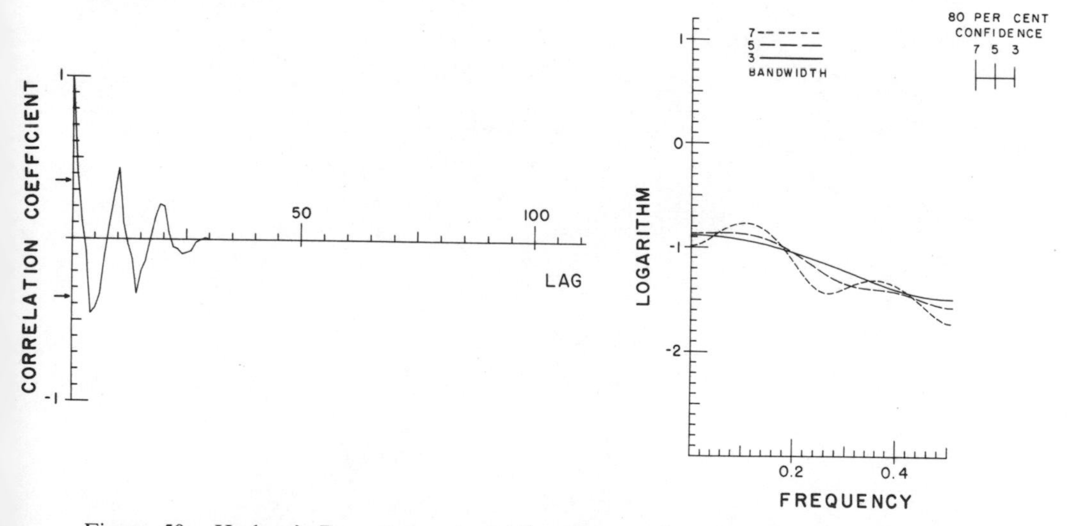

Figure 50. Hudson's Bay returns on wolf and coyote for all of Canada, 1879–1909, which may be predominantly coyote furs, demonstrate a clear cycle with a period of 10 years.

Prairie wolf or coyote, *Canis latrans*

Figure 51. Distribution of the coyote in Canada, based on Banfield (1974).

species. The wolf is essentially a big-game animal, although lynx, red fox, rabbits, muskrats, and other small game are an important diet item in some cases (Banfield, 1974). However, the wide range of food, combined with the wide range of distribution (fig. 53), might subdue or obliterate any local cycles when viewed on a national scale. On the other hand, the primary diet of the coyote is small mammals up to the size of woodchucks and rabbits (Banfield, 1974), and a predator – prey relationship with cyclic showshoe hares might well produce a cyclic flux in the coyote population.

Lynx rufus

The bobcat, *Lynx rufus*, is a close relative of the Canada lynx that is confined to the southern part of North America (fig. 54), inhabiting swamps, woodlots, second-growth forests, rocky hillsides, and deserts. Most breeding occurs in March or April, with an average litter of two to three kittens born after a gestation period of 50 to 60 days (Banfield, 1974; Petersen, 1966).

The bobcat's diet is more varied than the lynx, although it includes mostly hares and cottontails; other food items are mice and rats, squirrels, other small mammals, some birds and livestock, and traces of a number of other things. The major predators are cougars, coyotes, and

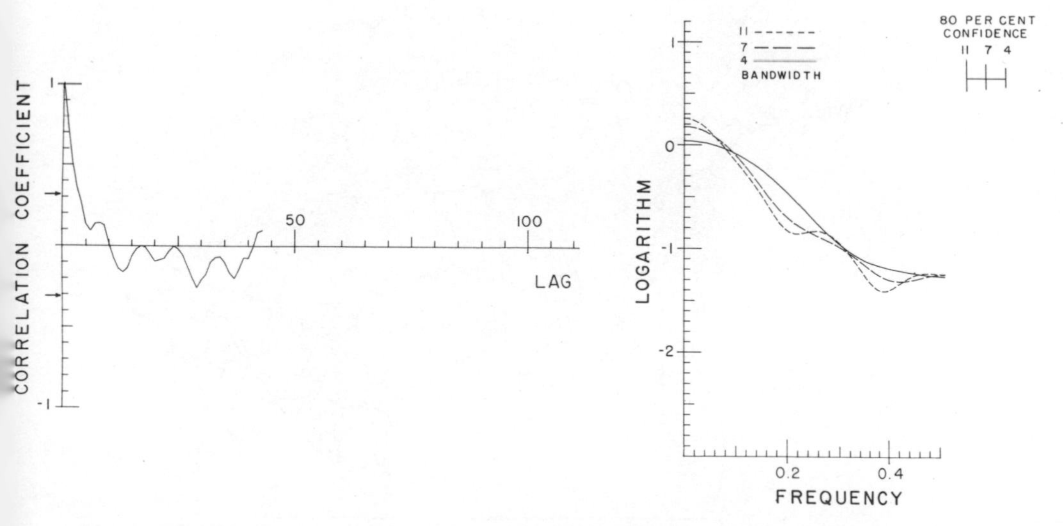

Figure 52. Time-series analysis of coyote fur returns from Saskatchewan between 1914 and 1957 support the contention that coyote may cycle with a 10-year periodicity.

Wolf, *Canis lupus*

Figure 53. Distribution of the wolf, *Canis lupus*, in Canada. (Based on Banfield, 1974.)

Bobcat or wildcat, *Lynx rufus*

Figure 54. *Lynx rufus*, or bobcat, distribution in Canada. (After Hall and Kelson, 1959.)

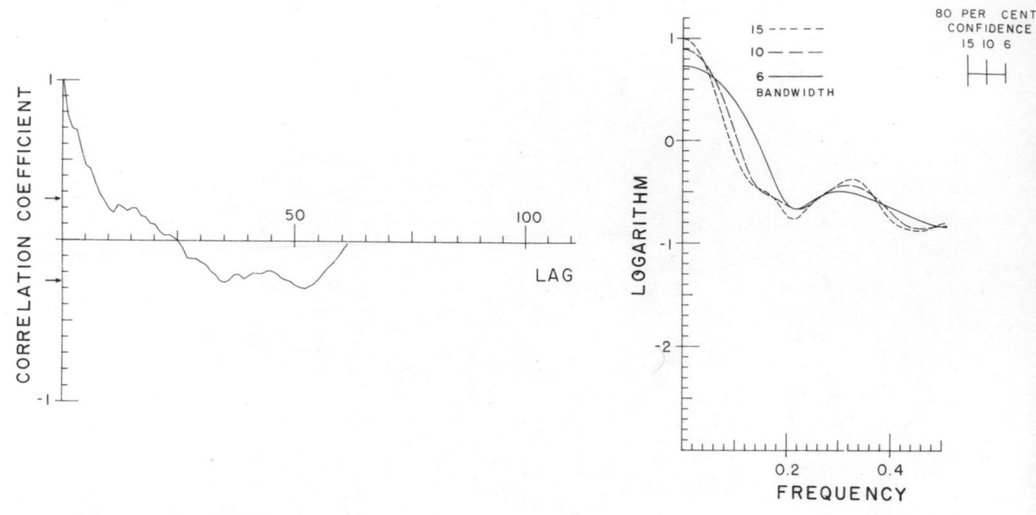

Figure 55. Bobcat populations in Canada, 1848–1909, do not reveal cycles when subjected to time-series analysis.

wolves, and great horned owls and foxes are important predators on the kittens.

Why, then, does the bobcat not cycle (fig. 55)? I would suggest that it is because this cousin of the lynx is a *facultative* consumer, and is therefore not confined to a narrow range of food supply in prey species; and because the habitat structure of the areas where the bobcat is resident is not conducive to cycles, a theory to be considered in chap. 4. In summary, the species listed in table 1 can probably be said to clearly evidence cycles.

TABLE 1

	Lagomorpha	*Rodentia*	*Carnivora*		
Cycle	*Leporidae*	*Muridae*	*Canidae*	*Mustelidae*	*Felidae*
4-year		*Lemmus* *Dicrostonyx* spp.	*Alopex lagopus* *Vulpes* spp.	*Martes americana* (?)	
10-year	*Lepus americana*	*Ondatra zibethicus*	*Alopex lagopus* *Vulpes* spp. *Canis latrans*	*Martes americana* *Martes pennanti* *Mustela vison* *Gulo luscus* *Mephitis mephitis*	*Lynx canadensis*

A Note on Phase Differences among Species

Not all populations evidencing 10- or 4-year fluctuations peak at the same time, and this fact has led to quite a bit of speculation on the meaning of phase differences among species. Bulmer (1975a), in an interestingly detailed analysis, determined phase differences for nine species by fitting data to a discrete-time model with sinusoidal oscillations and then calculating lags between each species and lynx (fig. 56). The data for red fox, marten, mink, muskrat, and skunk are from nineteenth-century Hudson's Bay Company data (from Jones, 1914); data for coyote, fisher, and wolverine are from Canadian Bureau of Statistics fur data since 1920 and represent averages of phase differences relative to lynx over all provinces where the given species is trapped in appreciable numbers. The

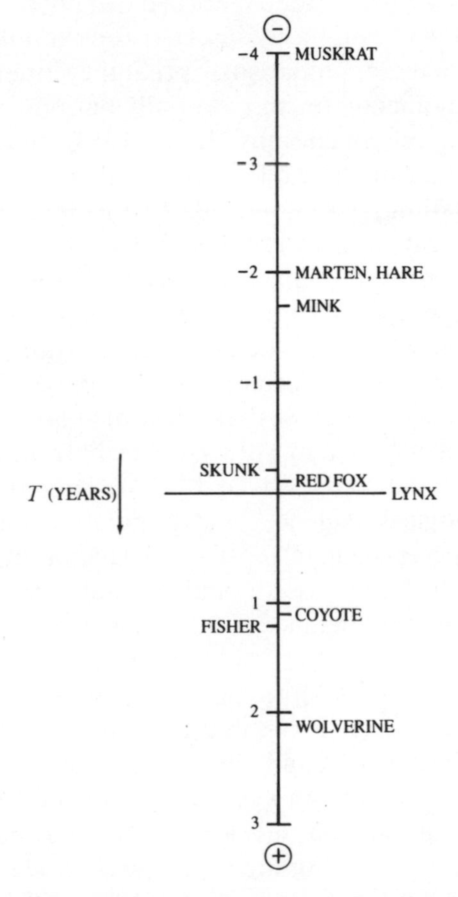

Figure 56. Phase relationships among nine species of small mammals. (After Bulmer, 1975a.)

results are comparable to what one finds by means of simple cross-correlation analysis (Finerty, unpublished). The position of snowshoe hare on the graph is suggested but not calculated, owing to lack of satisfactory comparable data (Bulmer, 1975a).

Bulmer used these phase differences to explain predation relationships among species: coyote, red fox, lynx, and wolverine all prey on hare and peak after hare, and this suggests that they may be tracking the hare cycle—the differences in time lags between species may be attributed to varying life histories, differing intensities of density dependence, or similar factors. Bulmer suggested that the muskrat cycle may be due to predation by mink, and that the mink and marten cycles may be attributed to cycles in horned owls and perhaps lynx and fisher, which are predators of snowshoe hares. Skunk cycles were tentatively attributed to skunk predation on ruffed grouse, although horned owl predation seems a more likely connection for the primarily insectivorous skunk.

The difficulty with these comparisons is that they attempt to represent a basically local phenomenon on an essentially nationwide scale, and this can lead to questionable conclusions. Butler (1953), dividing Canada into 63 sections and looking at phase relations in each from the subjective view of timing of population peaks, found that while lynx and fox generally peaked simultaneously, a large number of lynx peaks occurred 1 year before fox; and mink peaks, while generally occurring 1 or 2 years after muskrats, could with a significant frequency occur simultaneously with muskrats, and in the Mackenzie delta, mink tends to peak 3 to 4 years after muskrats. Preliminary spectral studies by Mark Boyce (1977, personal communication) suggest that muskrats may cycle in areas such as Louisiana, far outside the range of snowshoe hare and thus beyond the possibility of being indirectly driven by hare predators. The type of detailed local biological evidence (and population data) necessary to explicate phase differences is in most cases lacking (notably excepting Keith et al., 1977, to be discussed), and it would seem best to reserve drawing conclusions until detailed local studies have been pursued.

Concerning 4-year cycles, too few data are available to make statements beyond the facts that Moravian Missions data indicate a simultaneous flux in arctic and red foxes, that Labrador and Norway foxes are generally in phase before 1900 and about 2 years out of phase after 1900, and that foxes over most of Norway are not always synchronous. That local or neighboring populations are generally synchronous is not surprising, and I have as yet not found a clear *Zeitgeber* to demonstrate that the half-cycle difference between Labrador and Norway is anything but fortuitous.

4 *Causal Factors: Theories and Facts*

When we have the mathematicians showing us our business and making certain forecasts, it is time the biologists got together and showed whether the mathematicians are right or wrong.

Charles Elton, Matamek Conference on Biological Cycles, 1931c

In 1930, Charles Elton noted: "One of the features of the recent trend of biology has been a series of revolts against the practice of theorizing without sufficient facts, and such revolts undoubtedly result in a very beneficent feeling of exhilaration derived from the direct contact with raw facts, unhampered by the weight of dogmatic theories. At the same time there must come a point at which the new facts that have been collected are felt to be too raw and numerous, and it is at this point that the need for coordinating principles begins to be felt."

Today the situation often seems reversed, and we find ourselves buried in theories, unhampered by the weight of facts! Clearly, "the dangers involved in constructing an a priori model of a biological situation are apparent, but we must not go to the opposite extreme and conclude that such methods are valueless. Not only may they suggest further lines of experimental investigation, but, if we give up hope of constructing valid quantitative theories, ecology will remain a collection of isolated facts" (Moran, 1950a). Theories can often be viewed as types of facts in themselves, and the need for coordinating principles justifies exploring any theories that might help develop an understanding of the driving forces behind population cycles.

Most theoretical models are not meant to be literal descriptions of a biological microcosm, and one should not demand that prediction and reality coincide exactly—we clearly are not that knowledgeable about most ecological systems. Models can, however, lead to qualitative predictions which are "robust" to changes in many parameters, while being critically dependent on others, and this can lead to new, or reaffirmed, directions for research. Essentially, we are seeking organizing forms via mathematics to help bring our thoughts together and bring insights into possible mechanisms that may account for our observations.

Mathematical models in ecology generally describe numerical responses which, in a predation system, refer to changes in prey density in response to changes in predation intensity. It is clearly not sufficient for a model to generate numbers or patterns comparable to field records: the model must also reflect, or be unaffected by, observed functional responses in predator and prey, responses that represent changes in numbers of prey consumed by individual predators. When trying to ascertain possible causal mechanisms for such curious phenomena as cyclic fluctuations, the relationship between functional and numerical responses, a distinction introduced into ecology by Solomon (1949), would seem to be the sine qua non for any model.

In a system where predation occurs, a large number of variables could potentially affect prey mortality. Leopold (1933) arranged these in five major categories:

1. Prey density.
2. Predator density.
3. Prey characteristics (e.g., reactions to predators, intrinsic responses to prey density, physiological rhythms, etc.).
4. Density and quality of alternative food sources for predators.
5. Predator characteristics (e.g., food preferences, attack efficiency, etc.).

Most models concentrate exclusively on categories 1 and 2, since these are the numerical responses that have been recorded and that need to be explained. However, many proposed causal mechanisms fit into the other categories, and a reflection of this fact should somehow be acknowledged in any model; if these variables are omitted, the omission should be justifiable by demonstrating that the variables are constant, or clearly minor.

Another question that should be addressed is one of scale: Will cycles occur in populations of any size, or is there a minimum density, or territory size, below which cycles will either be unnoticeable or not occur? The idea that cycles are continent-wide in Canada is common. Cross (1940), however, working with red fox (*Vulpes fulva*) in Ontario, suggested that fluctuations are regional phenomena arising from local conditions, and a lack of regional synchrony in brown lemmings near Barrow, Alaska, implies the same (Pitelka, 1973). That a similar regionalism exists in Norway is implied by the fact that "lemming years" occur irregularly locally, although they occur somewhere in Norway about every 4 years

(Elton, 1942). Additionally, the theory that populations may be subject to epicentric dispersal (Elton and Nicholson, 1942a) implies that a more restricted territory size may exhibit fluctuations producing dispersal into neighboring territories. These facts should be clarified and acknowledged to be certain that they do not invalidate the model.

Tundra and Boreal Forest: Where Cycles Occur

The two major life zones of significance in discussions of cyclic population fluctuations are the tundra and the boreal forest (or taiga). Associated with tundra areas are short cycles with an average period of 3 to 4 years, and associated with coniferous forest areas are longer cycles with an average period of 10 years. Additionally, cyclic fluctuations have only been observed in the northern hemisphere. In this study, attention is specifically directed toward mammals in Canada and Norway because of the availability of long-term population data for these areas; whether or not the results will prove to be generalizable waits to be seen.

Is there any pattern in the climate or physical structure of these areas that could suggest why they evidence cyclic fluctuations while others do not? First, consider the tundra areas, where short-term cyclic population fluctuations have been observed in lemmings, grouse, and their predators. The word tundra is a collective term that is applied to the vegetation of both arctic and alpine regions, because the physical features of arctic and alpine vegetation are broadly comparable. Tundra regions are distinguished by the regular occurrence of long, very cold winters and short, cool summers, with little precipitation in any season. The flora are predominantly low-growing, forming a dense vegetation mat on the damp, cold ground. Trees cannot grow because only a thin layer of soil is released from permafrost in the summer and large plants cannot be physically supported. Moreover, the soil is cold and waterlogged throughout the summer and the growing season is very cool and very brief.

The tundra can be divided into four broad regional areas running from north to south: (1) rocky barrens, which have an incomplete plant cover; (2) moss and lichen heaths; (3) shrub tundra; and (4) wooded tundra, which merges into the boreal forest. Bogs and marshes are common in low areas where drainage is poor, and these are considered to be subclimax vegetation regions, since their margins are actively invaded by other plant species. Plant associations in the tundra are considered to be generally unstable (Rumney, 1968).

The climatic environments of arctic and alpine tundra are similar with respect to mean annual temperature and duration of snow cover, but may significantly vary in terms of insolation (high-latitude areas may experience as much as six continuous months with virtually no solar radiation, while many alpine areas experience regular diurnal and seasonal patterns); wind (the low Arctic experiences standard wind patterns corresponding to latitude, whereas many alpine sites experience extreme winds); and snowfall (often low in the Arctic and often extremely high in alpine areas) (Barry and Ives, 1974). In addition, Arctic tundra areas tend to be broadly continuous, whereas alpine areas are markedly noncontiguous and heterogeneous, owing to slope, orientation, and other factors affecting flora and fauna (Barry and Ives, 1974), and thus alpine areas exhibit a patchy "island" distribution (Hoffmann, 1974), which may prove to provide circumstances appropriate for dispersal and colonization, which in turn may be important in generating or enhancing cycles.

The sea and the mountains exercise a strong modifying influence over tundra areas. The sea renders the annual temperature variation of shore areas more moderate than that of areas more remote from the waters of the Atlantic or Pacific. In combination with heavier snowfall, it also retards the seasons in eastern North America and western Europe, making August the warmest month instead of July. Mountains can form an effective barrier against the warm air currents off the unfrozen surfaces of the ocean.

Winter temperatures in the European tundra are not as low as in the North American zone. The mean daily temperature range in North America is not much less in summer than in winter, whereas in the Eurasian tundra the summer range is about half that of the winter, except at marine stations in Norway (Rumney, 1968).

In Scandinavia the arctic tundra runs across the extreme northern part, reaching its most westerly extension in northern Lapland (fig. 57). This region has the same basic forms of mammals and birds as the North American Arctic, and represents an ecological area that essentially circumscribes the northern hemisphere (Elton, 1942). The factor which is peculiar to Scandinavia is that this arctic community of plants and animals extends far to the south, "down the complex and broken mountain range that forms the watershed between Norway and Sweden" (Elton, 1942). What begins as the broad continuous Arctic zone stretches southward toward the subalpine zone, or forest-tundra ecotone, which is generally a mosaic of low willow and dwarf birch or juniper, grasses, sedges, mosses, and lichens. Above the tree line is the alpine zone, which is

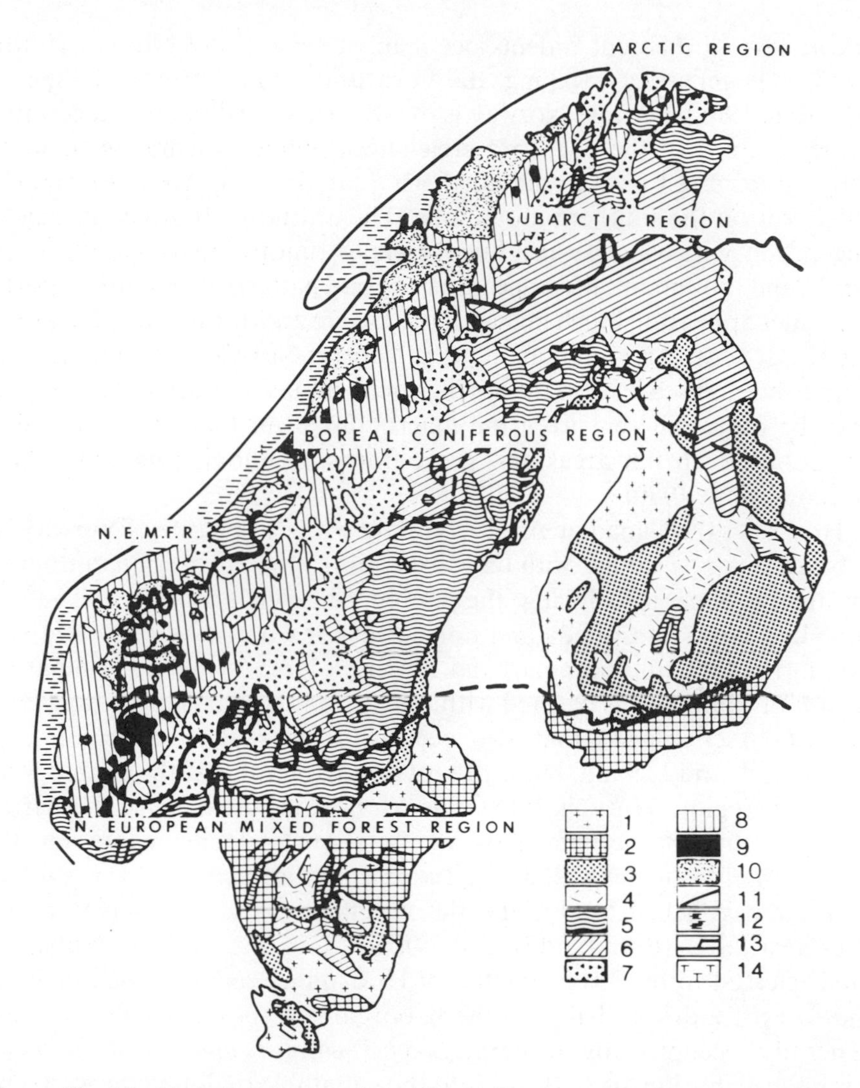

Figure 57. Morphological type regions of Scandinavia. 1, Plain, below 20 m high; 2, fissure-valley landscape; 3–5, undulating hilly landscape: 20 to 50 m high (3), 50 to 100 m high (4), and above 100 m (5); 6, monadnock plain (single remnant of a former highland rising as an isolated rock mass above a plain); 7, premontane region; 8–10, mountains: general fjells (8), fjells with plateaux (9), fjells with alpine relief (10); 11, fjord coast; 12, strandflat; 13, major fault; 14, table mountains. (Generalized from Rudberg, 1968.)

a basically barren area. The lemmings that form the famous migrations in Norway are chiefly from the subalpine area (Elton, 1942), inhabiting the drier parts, where they hide during the day, emerging at night to feed on grass and other plants. The lemming is the only rodent in the Norwegian subalpine area, but also partly inhabits open tundra reaching to the Arctic coast; several different rodents occur in other areas of Norway (Elton, 1942), but generally speaking, the flora and fauna are poor in species (Hustich, 1968). Because Norway is mostly arctic and mountain country, nearly a third of the country's area offers potential lemming habitats, ranging from sea level in the northern Scandinavian Arctic to 3,000 to 4,000 feet on the southern mountains. The altitudinally zoned mosaic of vegetation with vague transitions between tundra, mixed forest, grasslands, and mountains also appears as a broad pattern from south to north, the zones appearing much higher in the Dovre and Jotunheim plateaux in the south-central region than in north-central Norway. This merges into pure tundra at sea level in the north. Local climates in a particular region may be quite different from the macroclimate in that area because of variations from flat areas and great fiords, to mountains and valleys (Johannessen, 1970).

How does the Canadian Arctic compare with this picture of Norway? It is basically quite similar with three exceptions: the Norwegian altitudinal gradient is generally missing; the permafrost, which considerably affects the plant cover of Canada and northernmost Eurasia, is comparatively rare in Lapland and Finnmark (far northern Norway); and the subarctic region of Finnmark is covered with forests of low sparse birch over large areas (a phenomenon that also occurs in parts of northern Finland, Greenland, and Iceland, but is otherwise uncommon in the circumpolar subarctic region; Hustich, 1968). The Arctic area, often called the Barren Grounds (fig. 58), occupies most of the shore of Hudson Bay and the Ungava Peninsula, as well as the Labrador coast. This is dwarf shrub–sedge–moss–lichen tundra and the vegetation is generally continuous over a vast area (Barry and Ives, 1974). South of this region is a subarctic area referred to by Seton (1909) as a Hudsonian region, which includes the forest-tundra and the northern boreal forest. These are the areas generally occupied by lemmings, specifically *Lemmus trimucronatus*, whose distribution also extends into the subalpine region in the west, and *Dicrostonyx groenlandicus*. There is a noteworthy exception to this generalization, which is *Dicrostonyx hudsonius*, a lemming species confined to the Labrador–Ungava area of eastern Canada, and the only lemming in that area (Hall and Kelson, 1959). This is the lemming for which there are more observations of violent predator fluctuations (reflected in the fox trapping records from the Moravian Missions in Labrador) than for any other North American lemming. This seems even more interesting since the Labrador–Ungava region forms a distinct and extensive subarctic ecological subzone consisting of a mosaic of small islands of tundra and taiga ecosystems stretching southward to the

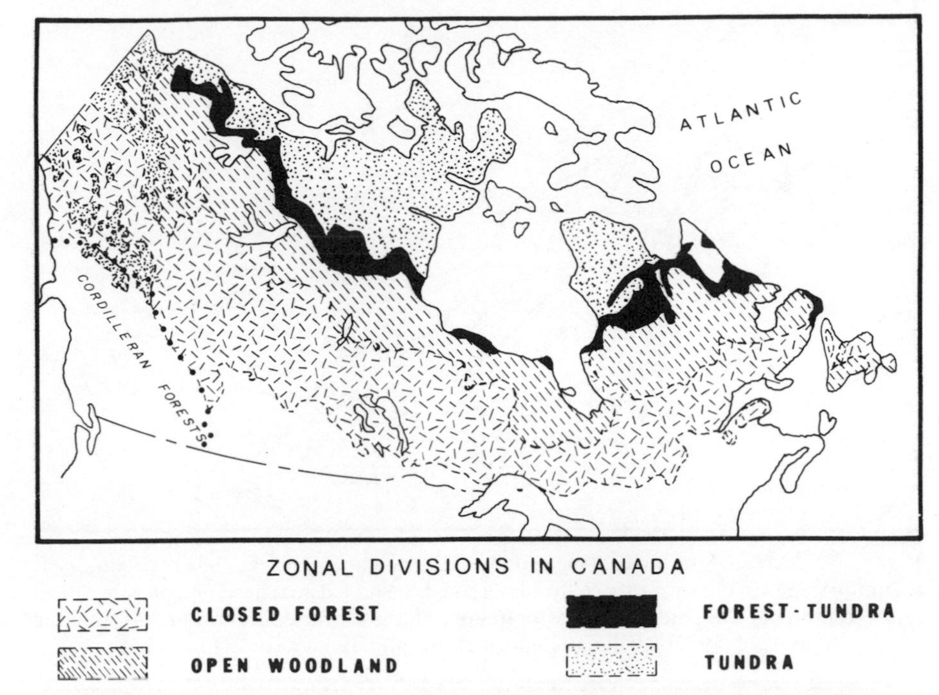

Figure 58. Zonal divisions in Canada. (After Hare and Ritchie, 1972.)

coniferous forest (Barry and Ives, 1974). This would seem to be a horizontal variation on the vertical Norwegian theme of a patchy "island" distribution, a fact that may help clarify some observations on cycles.

Coniferous trees and shrubs dominate the boreal forest, which spreads hugely across Canada and offers, in its southern regions, a home for the Canadian subfauna (fig. 58). Summers are short and warm, winters long and cold, and the seasonal transitions are rapid. Typically, the temperature is below freezing for about 6 months, and the annual precipitation is slightly more than for the tundra. Four basic types of landscape can be defined: (1) deciduous thickets which protect the coniferous seedlings from direct sun, and are transitory; (2) swamps, bogs, and marshes formed because of poor drainage of soil in the cold north; (3) open lichen woodland (forest-tundra), which blends into the tundra; and (4) close-crowned forests (Rumney, 1968).

In the North American boreal forest the frost-free period is about 60 to 110 days, compared to fewer than 50 days over the entire Arctic area (Hare and Hay, 1974). Northward-flowing streams (e.g., the Mackenzie River, whose headwaters lie about 15° south of its delta) locally moderate the climates of their lower valleys by transporting warmer waters from milder

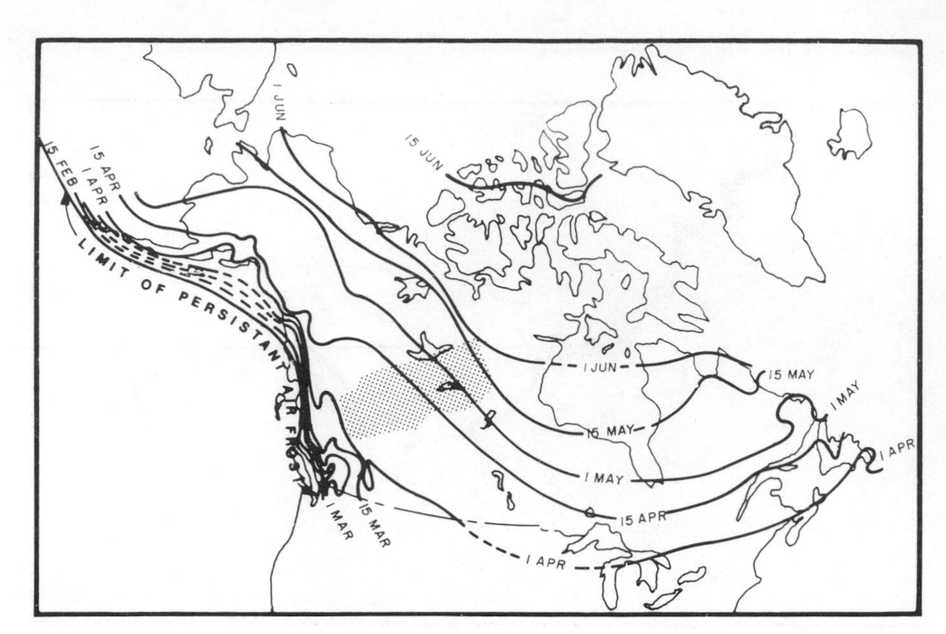

Figure 59. Mean date of rise of mean daily air temperature to 0°C, 1931–60. Note early warming trends in the west, especially inland Alaska. Shaded area here, and on subsequent maps, represents the Hudson's Bay Company Athabasca fur collection area (Elton and Nicholson, 1942b). (Climatic information after Bryson and Hare, 1974.)

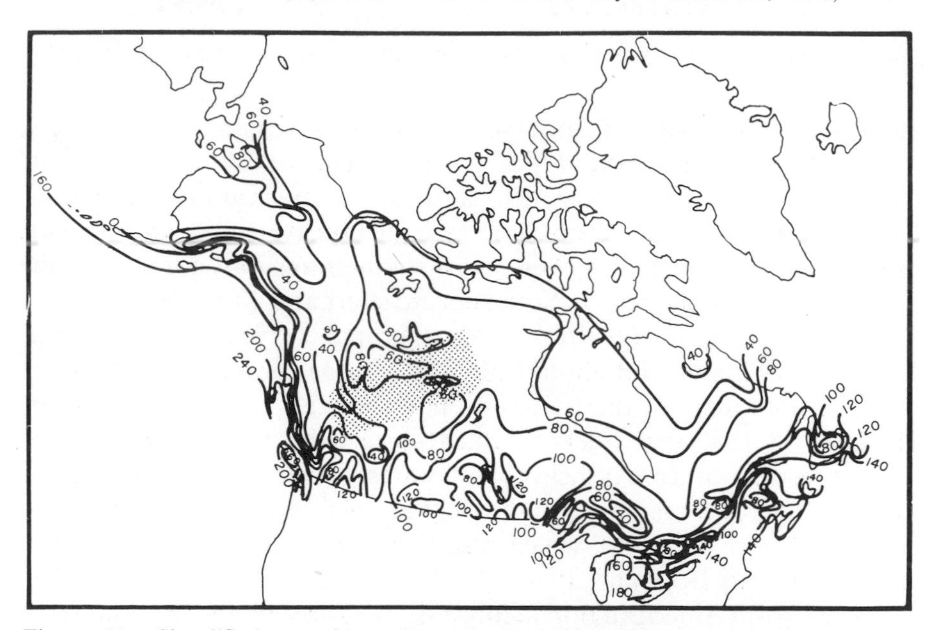

Figure 60. Simplified map of mean length in days of frost-free period, not standardized as to period. Note moderated climate of Athabasca area (shaded). (After Bryson and Hare, 1974.)

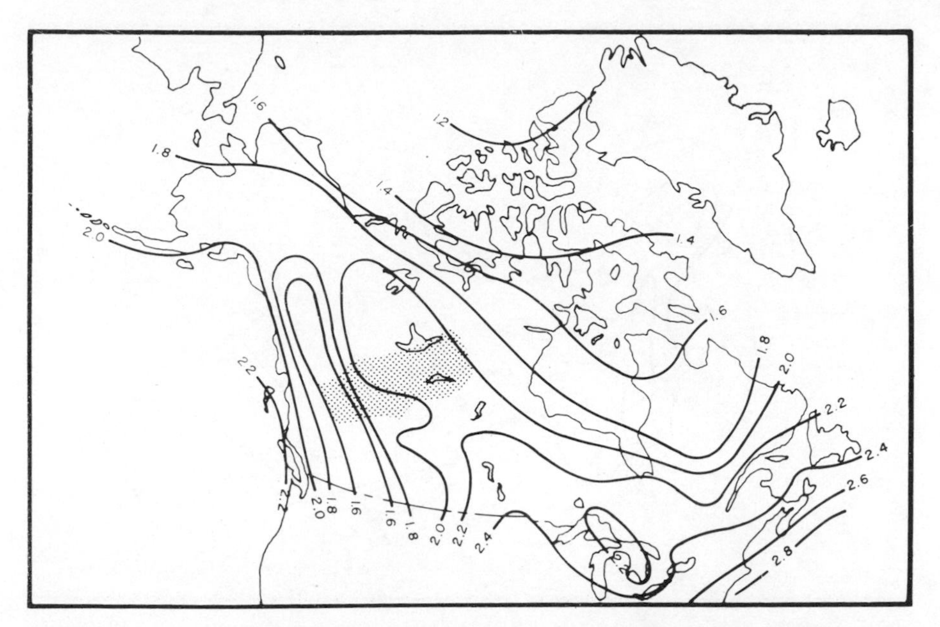

Figure 61. Mean precipitable water (cm) in July, 1957–64. Note higher precipitation in shaded Athabasca area. (After Bryson and Hare, 1974.)

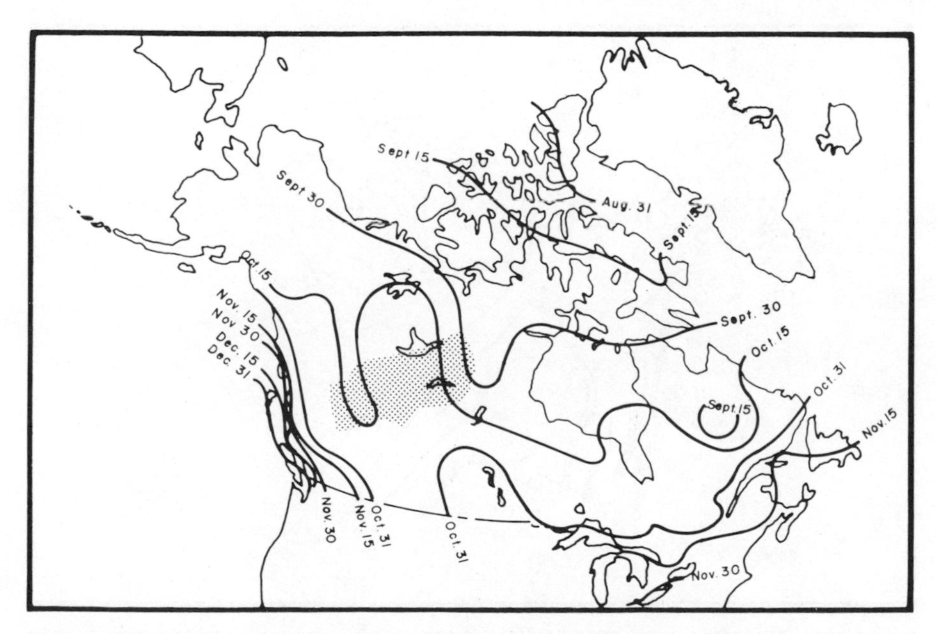

Figure 62. Median first day of snow cover greater than 2.5 cm. Note extended warm autumn season in Athabasca area (shaded). (After Bryson and Hare, 1974.)

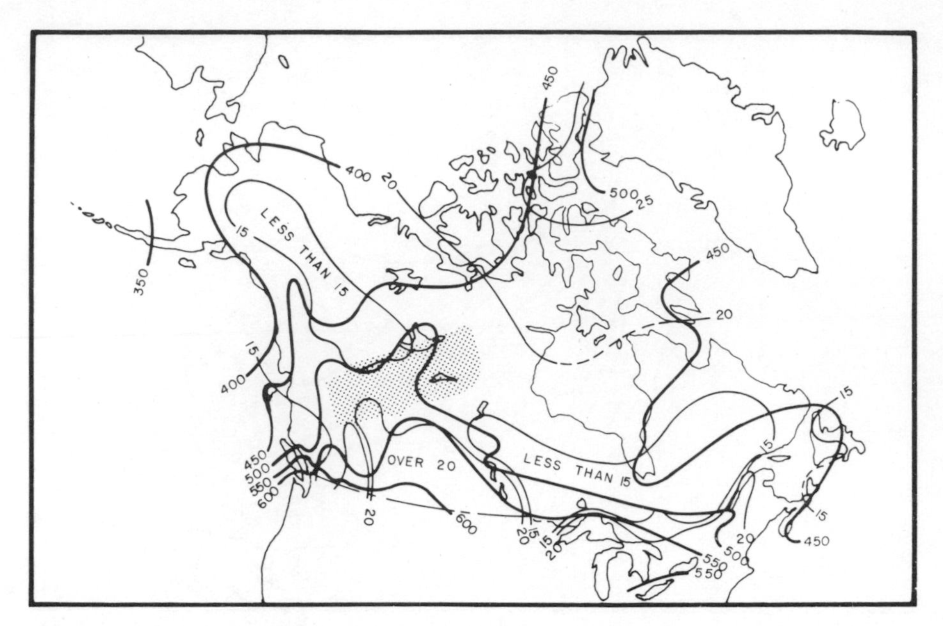

Figure 63. Mean global solar radiation in July, showing higher radiation in Athabasca area (shaded). (After Bryson and Hare, 1974.)

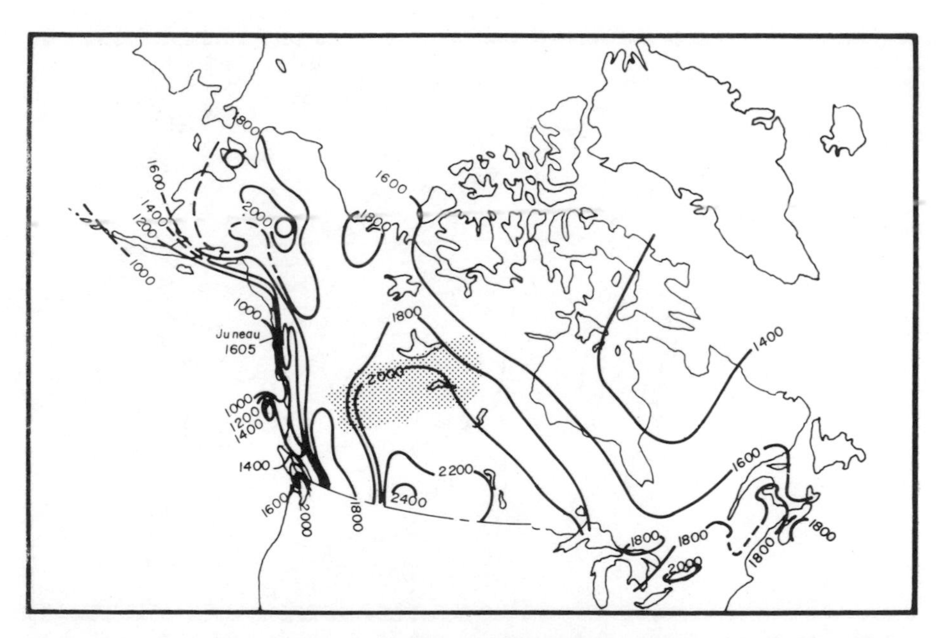

Figure 64. Annual mean hours of bright sunshine between 1931 and 1960. Note higher means in Athabasca area (shaded). (After Bryson and Hare, 1974.)

areas, causing fingers of forest to reach into the tundra. In northern Labrador and Ungava the forest merges unevenly with the tundra. Precipitation generally increases from the northwest to the southwest because of the northwestern mountain range. Tracks of cyclonic storms from the southwest in North America cause more abundant precipitation in the eastern forest, especially in terms of the number of days with snow (Rumney, 1968). Although July is usually warm, the wide range of summer temperatures means an ever-present threat of frost.

Several anomalies that occur in the Canadian forest climate regime are worth noting. For example, the spring thaw moves slowly and persistently northward until the beginning of April (fig. 59), but then an asymmetry appears between the heavily snow-covered east and the west: in the prairies and the Athabasca–Mackenzie basin the thaw travels northward rapidly. When the thaw approaches the Arctic tree line, however, it is retarded for some time, and this makes June the month of general thaw for all of the true Arctic (Hare and Hay, 1974). In addition, the climates around Lake Athabasca and Great Slave Lake are moderated, extending the length of the frost-free period significantly (fig. 60). These areas also experience a higher precipitation in summer (fig. 61) and an extended warm autumn season (fig. 62), higher mean global solar radiation in July (fig. 63), and more annual mean hours of bright sunshine (fig. 64). This may make the forest areas of the Athabasca basin, and northern Alberta in general, better equipped to efficiently restore depleted forage after a herbivore population peak.

The distinction between patchy and uniform environments should be kept in mind, since this concept may be important in helping to understand the nature of cycles and population movements in general.

Intraspecific Density-Dependent Models

Most of the mathematical theory of population cycles divides itself naturally into two categories: (1) theories that view cycles as intrinsic to a single species and generally assume that predator cycles are tracking prey cycles; and (2) theories that view some form of interaction as the driving force for cycles, thus interpreting cycles as an interaction of the prey with its food supply, with competitors for common resources, or with predators.

Several facts argue against predation as the major force driving cycles. In the case of microtines, two species in the same area may often cycle in phase (Krebs, 1964a, for Alaska; Tast and Kalela, 1971, for Finnish

Lapland), but sometimes one species will decline several months before another (Krebs and Myers, 1974; Tast and Kalela, 1971). Krebs and Myers (1974) suggest that in some declines predation is not necessary to cause the decline phases, although it may contribute significantly to the rate of decline and the ultimate low population level (see also Pitelka, 1973). Thus while microtine predators clearly respond to abundance in prey populations, the presence of predators may not be prerequisites for the initiation of a population decline.

For the 10-year cycle, one of the most often misquoted "proofs" that snowshoe hare cycle independently of predators is from Elton and Nicholson (1942b): "We have unpublished notes suggesting that the introduced rabbits on Anticosti have developed a cycle corresponding to the mainland one." This statement was taken as "proof" because Anticosti Island did not have a population of lynx, and therefore no predator was there to drive the prey–predator oscillation. What actually occurred was that Elton had met a Mr. Salzman (who was, as Mr. Elton recalls, a game warden on Anticosti) at the Matamek Conference on Cycles in 1931. From him and others Elton pieced together the fact that rabbits introduced onto Anticosti in about 1904 had "developed a ten year cycle synchronizing with the mainland, but with a much smaller variation between top and bottom, owing probably to absence of lynx which normally drives them down to scarcity during periods of low numbers following disease" (letter to Mr. Townsend from C. Elton, 4/14/32; C. Elton, personal communication). The series of Snowshoe Rabbit Enquiry reports in the Canadian Field-Naturalist published from 1935 onward, based on information collected by Elton's Bureau of Animal Population at Oxford, received a few reports initially from Anticosti and Newfoundland (where hare was introduced, probably in the 1880s, and where there are lynx), but these reports did not continue and "were not particularly useful, except that the Anticosti hares seem to have remained in at any rate noticeable abundance after the crash on the mainland" (C. Elton, personal communication). Concluding, Elton (personal communication) noted:

> My own opinion is that there must be "cycles" on both islands, but that there is no *scientific proof* that they were synchronised with the mainland. But I find it odd that you think the concept questionable *per se*. You simply can't dismiss the facts about mainland Canada itself, where thousands of the populations are in effect island populations (many without lynx) through habitat distribution, and yet where there has been for over 150 years a "national" cycle. The peaks and troughs are not simultaneous, but they are synchronised in the sense that they have never got entirely out of step, and show a high degree of relationship in time/region (see our lynx paper analysis of this). To this extent the argument about Anticosti is not vital.

I would agree that the argument about Anticosti is not vital, but for an additional reason: the existence of other predator species feeding on snowshoe hares. At the same Matamek conference, Mr. Elton met "the chief poacher on Anticosti Island, who had taken 1200 coloured fox there the previous winter"! (C. Elton, personal communication). This implies that the hares are not cycling in a predator-free environment, so that the possible existence of cycles on Anticosti cannot per se be used as evidence for prey cycling independent of predators. The fact that hares did not evidence the extreme variations of the mainland may be due to the broader diet of the colored fox.

Several factors argue against predation mortality as the determining factor for cycles in snowshoe hares. Keith (1974), working on a hare population in north-central Alberta near Athabasca, concluded that "predator populations which have built up on the increasing number of hares can only account for a small fraction of mortality in peak hare populations. However, after the hare population has declined sharply for 1 or 2 years due to effects of winter food shortage, while predators have remained at about the same level or even increased slightly, the hare – predator ratio is reduced to a point where rates of predation become high (inverse-density-dependence) and predators are the dominant mortality factor." (Also see Nellis et al., 1972; Keith et al., 1977.) This implies that peak hare populations provide food far in excess of predator requirements and must therefore decrease significantly (presumably due to some intrinsic mechanism) before predators experience an adverse response. We can also note that the reproductive rate and juvenile growth rate of the hares declines before the peak winter of critical food shortage (Keith, 1974; Cary and Keith, 1979), suggesting that predation does not initiate the decline. Although predators are thus an important contributing factor to population extremes (compare to Elton's views for Anticosti), this pattern suggests that they may not be the driving force. The fact that colored foxes evidence a 4-year cycle on the Ungava Peninsula, where lemmings and mice are the predominant prey, and a 10-year cycle in the forest areas of Canada, where snowshoe hare provides a major food source (which also seems true for marten; Elton, 1942; although marten may be an alternative prey of lynx in the 10-year cycle and a predator in the 4-year cycle) also argues in terms of predators somehow tracking prey cycles. Since so many predators experience cycles of similar period length, a simple underlying cause, such as an independent prey cycle, might be the driving force. This is reinforced by Bulmer's (1976) observation that the sharp asymmetric peak of the lynx periodogram demonstrates a remarkably constant period but does not resemble the expected theoretical shape for a hare–lynx interaction; the shape of the periodogram is what one would expect assuming that lynx cycles are driven by hare cycles.

There is one curious phenomenon that does not quite fit the predator-follows-prey view: evidence from movements of colored fox is said to suggest that migrants moving from the forest zone into the tundra continue to follow the rhythm of their homeland initially, although they apparently adapt to the "new ecological pressures" and later display the 4-year fluctuation characteristic of their new breeding grounds (Butler, 1951, 1953). This is probably due to a continuing flow of predators away from a population that has exceeded the current carrying capacity of a forest environment. This would broaden the population peak in the forest zone by providing an alternative food source for the foxes, and the influx of migrants would be recorded in the tundra areas as a 10-year flux. As the forest zone forage is restored, the flux would diminish and the tundra records would evidence only the local 4-year flux.

One can expect that many of the causal factors producing cycles in prey will also affect predators, but it seems likely that the cycle is basically intrinsic to the prey species and is not being generated by interactions between vertebrate predator and prey. The observation that rabbits may be increasing while lynx are dying off (Butler, 1953) can be interpreted by postulating that the rabbit population may have been pushed so far below lynx requirements that it could increase and still not meet lynx needs; and the fact that arctic fox may decrease while lemming are at their peak (Butler, 1953) may reflect an observation of a local naturalist who might not have noted if fox had emigrated to neighboring areas due to competition pressures; thus these arguments do not seem to definitively weaken the predator-follows-prey thesis.

What, then, is the role of predation in these cycles? Keith and his coworkers (1977), working in central Alberta, found some long-awaited detail to help answer this question for the 10-year cycle (fig. 65). Examining predation responses of lynx, coyotes, great horned owls (*Bubo virginianis*), goshawks (*Accipiter gentilis*), and red-tailed hawks (*Buteo jamaicensis*) to a changing snowshoe hare population over a 10-year cycle, they observed that predation on hares was low during years of increasing peak populations, increased sharply after the hare population declined significantly, and peaked 2 to 3 years after the hare peak. Numerically predators declined well after the hare. Apparently, peak hare populations provide significantly more food than predators require, and therefore predators are not adversely affected until hare populations decrease markedly.

An especially useful facet of this study is the clear distinction between

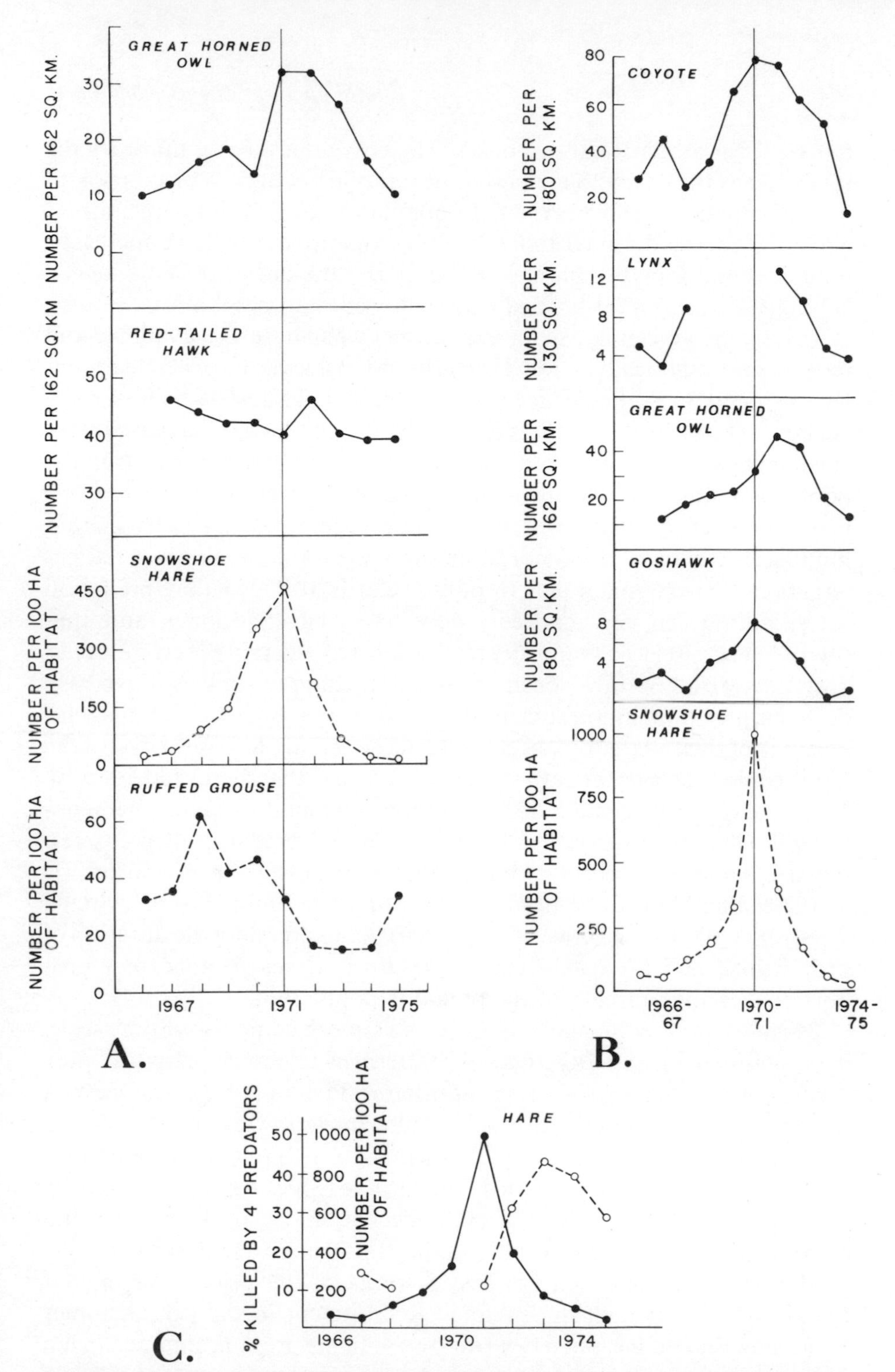

Figure 65. (A) Number of territorial great horned owls and red-tailed hawks near Rochester, Alberta, and their consumption of snowshoe hares during spring and early summer. Snowshoe hare and ruffed grouse densities for May 1. (B) Number of coyotes, lynx, great horned owls, and goshawks near Rochester over winter, and their consumption of snowshoe hares. Hare densities for December 1. (C) Winter predation rate on snowshoe hares in December. Solid line represents hare density, dashed line percentage of hares killed by four predators. (After Keith et al., 1977.)

functional and numerical responses. The coyote peaked in numbers the same year as the hare and increased consumption of hare in proportion to hare abundance, although the hare population decline was precipitous, while the coyote declined more slowly. Lynx appear to have increased hare consumption with hare availability, but the decline of lynx lagged behind the hare. Great horned owls followed hare populations closely, increasing the percentage of hare in the diet when hare was available and responding numerically parallel to the hare. Goshawks also increased, but lagging the hare in the decline phases. Numbers of goshawks increased, peaked, and declined parallel to hare. Red-tailed hawks increased hare consumption from approximately zero to nearly 50 percent, but the population was numerically relatively stable, reflecting this predator's broad base of alternative prey (and also reemphasizing the fact that numerical response may not accurately describe predator–prey interactions). These functional responses clarify the idea that predation rates continue to rise markedly *after* hares have declined, and that predation persists as a major depressive form of mortality for a time *after* prey have reached low densities. When predator numbers have also declined, prey populations escape this control, and, by rates of increase far in excess of their predators, begin to increase rapidly to levels beyond the food requirements of predator species. When another factor [Keith et al. (1977) suggest a possible hare–vegetation interaction] again intervenes to reduce the prey to levels below predator needs, predators again exert a massive depressive effect. This probably means that predation (1) increases fluctuation amplitude by reducing prey numbers to extreme lows, (2) lengthens the between-peak interval by extending the duration of prey nadirs, and (3) synchronizes prey fluctuations to an extent proportional to the mobility of the predator population.

Because lemming fluctuations occur on a more compressed time scale, it is difficult to assess whether the pattern of lemming predator–prey relationships exactly parallels that for hares. Pitelka (1973), who views a peak year as a high-population winter plus the following summer, found at Barrow, Alaska, that heavy avian predation on exposed lemmings (who, having massively grazed the grass–sedge–tundra, are easily visible to predators) began the cyclic decline in the peak summer and continued through the summer, ending in the postpeak winter. The predation was heaviest in the first 3 weeks of the peak summer, but remained high through the summer. Recruitment in the peak summer could slow the decline, but never reversed it. In the prepeak summer, avian predators paced the population increase: breeding pomarine jaegers (*Stercorarius pomarinus*) are especially important at this point.

The major predators on the lemmings at Barrow are pomarine jaegers, least weasel (*Mustela rixosa*), and snowy owl (*Nyctea scandiaca*). The arctic fox, which moves to the coast in autumn and winter, may also be important. The density of breeding pairs in jaegers is consistent in peak years, suggesting that numbers of this species are being limited by intraspecific territorial behavior (Pitelka, 1973). As with snowshoe hares, this type of predator population constraint may create a situation where prey density far exceeds predator requirements, although the presence of other predators in quantity may offset the jaeger's constraint. For example, while the breeding density of snowy owls may remain relatively constant during the increase and peak, many nonbreeding snowy owls may invade the area also. And while the breeding population of the snowy owl may not exhibit a numerical response, the average clutch size can increase three- to sixfold between a moderate and a peak year (Pitelka et al., 1955a). In addition, least weasels, which can prey directly on both nestling and adult lemmings, with predation persisting through the decline phase, appear to be at least as important as jaegers in decimating populations.

Krebs (1964a), working on brown collared lemmings at Baker Lake near Hudson Bay, found a population decline in the absence of predation, suggesting that predation may not be the controlling factor here. However, this observation was made for only one winter. If, as an example, an early thaw and refreeze had flooded the frozen vegetation not inhabited by lemmings, migrating avian predators might circle the territory and depart for want of sufficient prey. The high juvenile mortality in the summer of decline might also be due to some such major change weakening the population. Consequently, it seems difficult to assess the nature of this particular decline.

Most other predation studies are concerned with nonarctic microtines. However, a theory emerges from these studies (Pearson, 1971) which seems consistent with the observations at Barrow. This theory suggests that avian predators are not sufficiently intensive to determine trends in rodent populations, since they are absent during low prey abundance. Mammalian predators, however, being less mobile, may act to decrease the populations to very low levels. Predators may also act to prolong the period of low numbers. Unfortunately, little is known about the period of low numbers (Krebs and Myers, 1974), but this view of the role of predation in microtine cycles is quite similar to that discussed by Keith for snowshoe hare, the unresolved question in both cases being the initial cause(s) of population decline.

In summary, predation seems to play a large role in fluctuations of both snowshoe hare and lemmings, but it does not appear to be an isolated determining factor for either cycle. Working with the possibility that cycles might represent intraspecific density-dependent responses in the prey species, we can try to find a mathematical formulation.

The Verhulst–Pearl logistic equation

$$\frac{dN}{dt} = rN\left(1 - \frac{N}{K}\right)$$

represents a simple model that describes the basic features of a single species population in a limited environment (for an enlightening history of this important equation, see Hutchinson, 1978). Here N is the population size, r represents the intrinsic rate of increase of the population if removed from environmental limitations, and K is the equilibrium value of the population, or carrying capacity. The term $r(1 - N/K)$ is the density-dependent form of the per capita growth rate. It seems reasonable to assume that, despite the historical arguments concerning density-dependence (Lack, 1966), these populations do exhibit this type of regulation; Bulmer (1975b) has proposed a method for testing this and suggests density dependence for colored fox, arctic fox, lynx, mink, and muskrat in Canada. Mathematically, this simple logistic can be regarded as a two-term Taylor series expansion of more general density-dependent models (Lotka, 1925). The logistic equation represents a large class of population equations with regulatory mechanisms, called nonlinear, all of which have a stable equilibrium point (at $N = K$ for the logistic) with any disturbance tending to fade monotonically back to the equilibrium level (May, 1976b).

The logistic equation is highly simplified, and at least four of the underlying assumptions (Pielou, 1969) are clearly violated in northern cycles:

1. Environmental factors are probably variable enough to affect birth and death rates.
2. Population growth rate (especially for prey) may not be density-dependent at the lowest levels.
3. Death rates, for prey, are likely enhanced by factors like predation, and a time lag may be involved.
4. Populations may not maintain stable age distributions (e.g., for snowshoe hares, see Keith, 1974).

However, the mathematical difficulties of approaching many ecological models, either deterministic, stochastic, or deterministic with stochastic terms, has led ecologists to focus on such relatively simple deterministic models.

It was suggested by Hutchinson (1948) that the logistic equation might describe a predator–prey system more realistically if a time lag were included. The time lag might result from environmental effects, such as vegetation recovery time, from physiological factors within members of the population (e.g., changing reproductive rates; Schaffer and Tamarin, 1973), or from the time between birthrate depression due to high densities and the subsequent decrease in the adult population (one generation time; May, 1976b). One could consider a time lag to be a polynominal function of biological parameters which determine the time scale of a species' ability to respond to input of new information by changing its population growth characteristics. A short time lag would result in an oscillatory return to equilibrium, but a long time lag could produce violent and persistent oscillations between finite limits. The population function might even include a time lag so that, should a random factor push the population far from its equilibrium value, a slow return to equilibrium would result instead of a violent flux or crash. Since time lags were first suggested, an extensive literature has developed (e.g., consider reviews in Goel et al., 1971; May, 1973a).

Consider the delay-differential equation

$$\frac{dN}{dt} = rN\left(1 - \frac{N_{t-\tau}}{K}\right)$$

first used in economics over 40 years ago (Hutchinson, 1978), independently introduced into ecology by Hutchinson (1948), and detailed by Wangersky and Cunningham (1956, 1957a). Here τ represents inherent time lag involved in the operation of some density-dependent negative feedback mechanism (represented by the factor $1 - N/K$). This equation can be viewed as a basic representation of a large class of models in which the regulatory term may take on rather complex forms (for detailed examples, see May, 1973a). If the feedback time delay τ is larger than the natural response time of the system ($1/r$), the population will tend to produce oscillatory returns to equilibrium. Thus for

$$0 < r\tau < e^{-1}$$

this population will have a monotonically damped stable point; for

$$e^{-1} < r\tau < \tfrac{1}{2}\pi$$

it will have an oscillatorilly damped stable point; and for

$$r\tau > \tfrac{1}{2}\pi$$

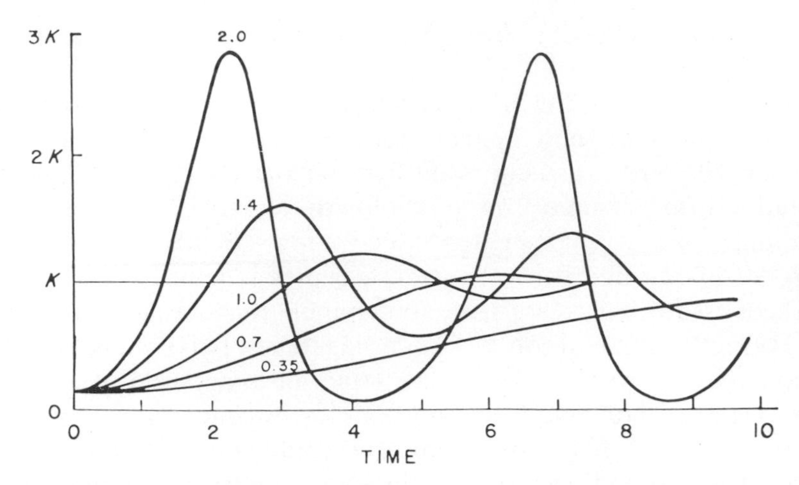

Figure 66. Ideal behavior of populations growing according to the logistic with a time lag τ, for various values of $r\tau$ (Cunningham). (From Hutchinson, 1978.)

it will have stable limit cycles. The amplitude and period of the cycle will be uniquely determined by the model parameters (fig. 66). The e^{-1} and $\frac{1}{2}\pi$ are specific for this model, but the idea of a stable equilibrium point giving way to stable cycles once the delay sufficiently exceeds the natural response time is characteristic of a large class of models incorporating time-lag regulation (May, 1976b).

It is especially important to observe that when stable limit cycles emerge in the described population, their period is approximately equal to 4τ, even when the population amplitude (maximum/minimum) changes by several orders of magnitude (see table 2; after May, 1976b). It is refreshing to know that the mathematics can demonstrate the kind of stability with respect to varying amplitude that is observed in nature.

TABLE 2

$r\tau$	*Amplitude* (N_{max}/N_{min})	*Period Length*
1.57 or less	1.00	—
1.6	2.56	4.03τ
1.7	5.76	4.09τ
1.8	11.6	4.18τ
1.9	22.2	4.29τ
2.0	42.3	4.40τ
2.1	84.1	4.54τ
2.2	178	4.71τ
2.3	408	4.90τ
2.4	1040	5.11τ
2.5	2930	5.36τ

Source: After May (1976b.)

Using this fact, May constructed a naive theoretical curve for τ slightly less than unity which evidences a 4-year periodicity. If the cycle is slightly different from 4 years, one might expect it to lock into an integral year due to the strongly seasonal environment, so that the average cycle might be 4 years. The problem of having similar period lengths for many species is obviated by the model's intrinsic period length of 4 years and the cuing of well-defined seasonal change, the only immediately obvious major requirement being to biologically justify a time lag slightly less than unity for all 4-year species.

When compared to Shelford's (1943) data on lemmings (*Dicrostonyx*) from the Churchill area in northeastern Manitoba on Hudson Bay (fig. 67), the fit of this admittedly crude model is quite convincing. However, several significant biological questions need to be considered before such a simple model can be accepted. One important question pertains to the constancy of r, the intrinsic rate of increase. This rate of increase will depend on a number of factors, including litter size, pregnancy rate, breeding season length, age at sexual maturity, sex ratio, mortality, and, since the cycles seem to represent at least initially local phenomena, dispersal. Studies at various other tundra sites have suggested that litter size, pregnancy rate, and sex ratio are not important factors determining cycles (Krebs and Myers, 1974). However, it seems likely that there are significant changes in the length of the breeding season during the cycle, with extended summer breeding and perhaps winter breeding during the increase phase, a shortened summer breeding season, and no winter breeding at the peak or during the decline, with a possible delay in the onset of summer reproduction during the decline. Age at sexual maturity may vary over the cycle, at least in some microtines; adult

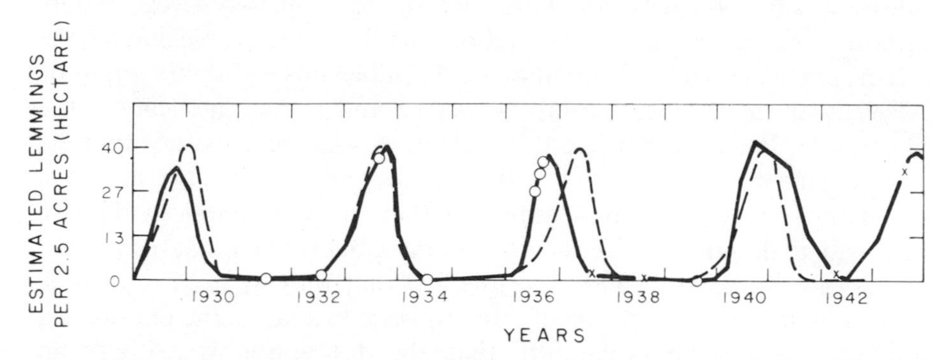

Figure 67. Shelford's (1943) data on lemming populations in the Churchill, Manitoba, area compared with May's simplified theoretical curve (dashed line) based on the simple time-delay logistic. The time delay is taken to be less than 1 year (0.72 year). (After May, 1976b.)

mortality rates are likely low in increase and peak phases, and high in the decline phases, while prenatal mortality does not seem affected by cyclic changes; and dispersal, although inadequately studied, may have a sizable effect on cyclic variations (Pitelka, 1973) and may produce both qualitative and quantitative effects on the population (see Krebs and Myers, 1974, for an excellent review of these and related aspects of population cycles in small mammals). If r is a multifactor, time-dependent polynomial function, it is difficult to know how this might affect the predictions of the model.

It is also possible that K, the carrying capacity or equilibrium population, is a time-dependent variable. If this time dependence is included in the logistic model, a population with a relatively large intrinsic growth rate r, hence short response time $1/r$, will tend to track periodic fluctuations in K, while a population with relatively small r will tend to average over basically all fluctuations in K if the response time is short or long with respect to the period of environmental flux (May, 1976b). (In stochastic models the effects of statistical variations in the environment may not average out, because the covariance function of the environmental random variables must be considered along with their means and variances; Kiester and Barakat, 1974.) Although it is conceivable that predators may be oscillating by tracking a cyclic carrying capacity (i.e., fluctuations in numbers of prey), the prey fluctuations remain to be explained. To date there has been no clear evidence of a cyclic environmental variable that prey species might be tracking. We will return to this problem when considering the two-species system.

The question of what biological mechanism(s) produce the time lag is recognizably unelucidated. Table 2 clearly shows that a large degree of allowable variation in τ can still produce an approximately 4-year period, although this neglects possible variations in r. It is conceivable that if r, the intrinsic rate of increase, is significantly variable, this variability might be offset by an inverse relationship between r and τ, the time delay in the negative feedback mechanism. It is, in fact, reasonable to assume that the negative feedback mechanism is density-dependent. Alas, the model is becoming somewhat complex! It may be that this simple model will prove adequate to describe what has been observed, but the fact that the data fit should not be allowed to obscure the fact that we may need to include quite a bit more biology to justify the suggested values of the parameters and to increase the probability that the description is actually an explanation.

Parenthetically, we might note that, since periodic population fluctuations can be generated so easily by introducing time lags, which are an inevitable component of any real ecological system, one can wonder why so few clear examples of periodic oscillations have been found. Although there appears to be strong selection against long time lags, it may be that certain species. like microtines and snowshoe hares, have found a way to make these useful (Hutchinson, 1975). However, a mechanism for the evolution of long-term time lags is difficult to imagine, and quite a bit more work will be required before such a theory can be verified or disproven.

Recent studies of the complicated dynamics of simple mathematical models (see, e.g., May, 1976a) again emphasizes the necessity of accurate, extensive biological information to evaluate causation. Consider the range of solutions of a simple model:

$$N_{t+1} = f(N_t)$$

If $f(N_t)$ is density-dependent (nonlinear), this is a nonlinear first-order difference equation. The function $f(N_t)$ will increase to some maximum monotonically and then decrease monotonically beyond that maximum. The function will usually contain one or more parameters which will determine the severity of this nonlinear behavior. This might be a description of a seasonally breeding population with nonoverlapping generations.

For illustration, consider the logistic difference equation

$$N_{t+1} = rN_t(1 - N_t)$$

for $0 \leq N_t \leq 1$. [If $N_t > 1$, subsequent iterations go to infinity, meaning that the population becomes extinct (May, 1976a).] For $r > 4$, N_{t+1} may not be in the unit interval; for $0 \leq r \leq 1$, $N_t = 0$ is the only static state (Hoppensteadt, 1976). For $1 < r < 3$, the stable static state $N_t = (r - 1)/r$ is present. For $3 \leq r \leq 4$, the behavior becomes complex and enters a region termed "chaotic" (May, 1976a). As the parameter r passes the value $r = 3$, the basic equilibrium value of N_t becomes unstable, and at the point where this occurs two new and initially stable equilibrium points emerge, between which the system oscillates in a stable cycle of period 2; this phenomenon is called bifurcation (fig. 68). As the parameter r continues to increase, the period 2 points become unstable and themselves bifurcate to give a cycle of period 4, which is initially stable. For r increasing even further, this gives way to cycles of period 8, 16, . . ., 2^n. This produces an infinite sequence of cycles with period 2^n ($n \to \infty$), but the process is convergent, being bounded above by $r = 3.57$. Beyond this accumulation point there are an infinite number of cycles of different periods. Sarkovskii (see Hoppensteadt, 1976) showed that the reverse of this sequence of

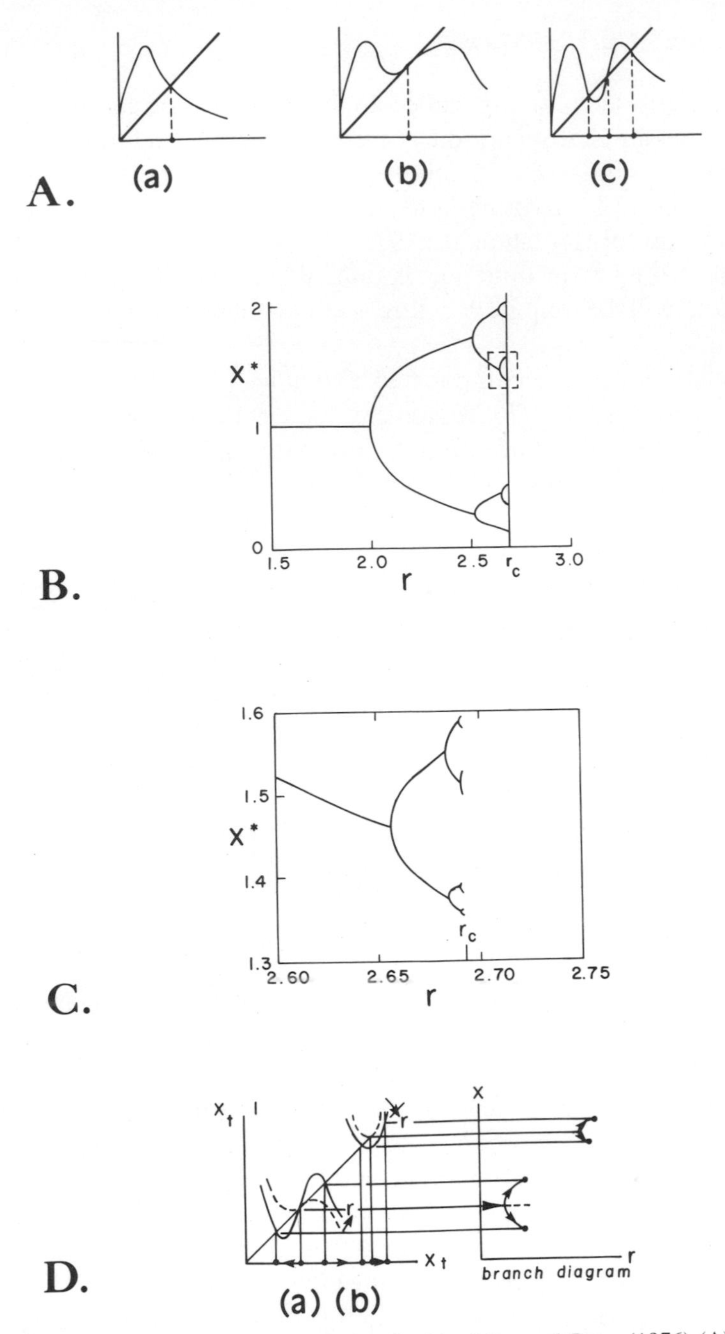

Figure 68. The bifurcation process as described by May and Oster (1976). (A) For $r < 2$, the original fixed point is stable (*a*); for $r = 2$, the original fixed point bifurcates (*b*); for $r > 2$, two period-two attracting fixed points emerge and the original fixed point becomes a repellor. (B) Hierarchy of stable fixed points, each arising by bifurcation when a previous fixed point becomes unstable, emerges as the parameter r of the logistic difference equation increases. This sequence of stable cycles of period 2^n is bounded by the parameter value r_c. (C) Detail of the bifurcation process in the indicated box in (B). (D) Bifurcation for $r < -1$ (figure a) and $r > +1$ (figure b).

period lengths for emergent cycles is $3, 5, 7, 9, \ldots, 3.2, 5.2, \ldots, 3.2^2, \ldots, 3.2^3, 5.2^3, \ldots, \ldots, 2^{n+1}, 2^n, \ldots, 2, 1$, and that if the period p in this sequence is a solution for this equation, all periods to the right of p in this list are also solutions. Beyond period 3 (at $r = 3.8284$; May, 1976b) there are cycles of every integer period plus an uncountable number of aperiodic solutions. A smooth function such as the logistic difference equation, however, has, for any specific r value, one stable unique cycle that attracts basically all initial points (May, 1976a; Guckenheimer et al., 1977).

The phenomenon of bifurcation is general for many difference equations used in the biological literature, and is also characteristic of continuous-time models (May and Oster, 1976), which are more appropriate for cycles where seasonal breeding and overlapping generations occur in most species involved. This implies that it may be virtually impossible in many cases to distinguish data generated by a simple deterministic process from stochastic noise.

To make matters more interesting, if one investigates the pattern of bifurcations away from the equilibrium point for some density-dependent population models, there may emerge a "strange attractor," an attracting limit set which is neither an equilibrium point nor a periodic orbit. Under certain circumstances, however, the strange attractor may exhibit the features of a periodic orbit with some superimposed randomness (Guckenheimer et al., 1977). Consequently, it may be virtually impossible to differentiate, on the basis of population numbers alone, between a deterministic cycle with stochastic perturbations and a purely deterministic strange attractor, which might falsely appear to be periodic. By rejecting models with strange attractors one essentially asserts that none of the observed irregularities are generated by internal (density-dependent) population processes. Only our biological knowledge of population dynamics can begin to direct us to causes underlying population fluctuations, and even then our present mathematical understanding will have to advance before general assertions can be made.

Two Interacting Populations: Herbivore–Forage Relationships

The problem of distinguishing useful models of single-species cyclic population dynamics is sizable, but models of two interacting populations may point toward relevant explanations for cycles. The biological facts argue against a carnivore–herbivore predator–prey oscillation, but the possibility of a herbivore–plant oscillation has not been excluded. The herbivore might deplete its food supply, either by decimating the plant cover or by exhausting those plants which are nutritionally important to the species, and subsequently undergo a population decline due to malnutrition or starvation. There is the additional possibility, especially in the tundra, that destruction of the ground cover might remove enough protective vegetation to visibly expose herbivores to increased predation.

Food Shortage in Lemming Cycles

Several investigators have suggested food shortage as a possible explanation for the population decline in Canadian lemmings (Elton, 1942; Pitelka et al., 1955b; Pitelka, 1959). Lack (1954b) thought the lemming cycle might be connected to overexploitation of the habitat by lemmings, with concomitant destruction of cover, resulting in greater exposure to predation. Pitelka (1957) envisioned the lack of food produced by high lemming density as leading to malnutrition, reduced reproduction, and a subsequent population decline.

It seems most likely that the "lack of food" might actually be the prolonged lack of a specific nutrient rather than visible elimination of forage (Hutchinson and Deevey, 1949; Lauckhart, 1957; Curry-Lindahl, 1962; Pitelka, 1964; Schultz, 1964; Kalela, in Clough, 1965). That lack of a specific nutrient can affect populations has been suggested for moose populations of Isle Royale, Lake Superior (Botkin et al., 1973), where populations may be limited by the quantity of sodium that can be collected during the brief summer season of aquatic grazing; Harper (1977) noted that the "life cycle strategy is likely in such a case to be influenced by the optimal allocation of sodium between parents and offspring and between the various sodium-demanding activities."

This idea of nutrient deficiency may account for Pitelka's (1973) observation that, after 6 years of low numbers of brown lemmings (*Lemmus*), an unprecedented increase of collared lemmings (*Dicrostonyx*) occurred in the Barrow study area. The periodic mowing of the tundra vegetation mat by *Lemmus* may not allow for sizable expansion of the *Dicrostonyx* population either due to lack of food or insufficient time for recycling of some nutrient necessary to *Dicrostonyx*—this may be a phenomenon peculiar to climatic patterns in certain tundra areas such as Barrow. If a different nutrient required by *Lemmus* is not available, sufficient time may pass to allow nutrient-rich forage for *Dicrostonyx* that remains unsuited to *Lemmus*. Alternatively, *Lemmus* may be the only nutrient-restricted species of the two, and may normally eliminate *Dicrostonyx* by competition, as evidenced by the decrease in *Dicrostonyx* during the winter ingress of brown lemmings in 1970–71 (Pitelka, 1973). A similar situation may occur in Norway, where the lemming population appears generally highest in the year when the vole population has crashed (Wildhagen, 1953).

There are several ways in which nutrient quality of forage plants might be depleted. Defoliation removes nutrients from plants, which can be especially important in perennial species in an area such as the tundra, where 65–71 percent of the total biomass is living matter and where nutrients are limited (Kimmins, 1970). Defoliation may also affect nutrient supply indirectly by removing the soil insulation and allowing more of the soil layer to thaw; it has been hypothesized that uptake of minerals by plant roots confined to the top nutrient-rich layers of soil may be greater than when roots occupy a larger, more dilute zone (Schultz, 1964). Another source of depletion, which could be important in the coniferous forest, is production of a heavy seed crop, which severely depletes soil nutrients locally (Svardson, 1957). These mast years can also affect mosses, since a large part of the moss nutrient supply comes in the form of chemicals leached from the canopy above (Kimmins, 1970). Nutrients are restored to the system in the urine and feces of defoliators and by decomposition of litter, but full restoration may take 2 to 3 years in the tundra, and even longer in the boreal forest, owing to climatological restrictions and variations in the chemical composition of the litter (Kimmins, 1970).

To establish the relationship of a herbivore to its food supply, three basic questions should be addressed (Krebs and Myers, 1974): (1) Do the herbivores exhibit clear food preferences in their choices? (2) How does herbivore foraging affect the habitat? (3) Is there evidence that food quality and/or quantity are limiting for increasing populations?

Studies of food habits demonstrate that fresh green material is preferred by lemmings summer and winter (Kalela, 1962; Kalela and Koponen, 1971; Melchior, 1972; Speller, 1972) with *Lemmus* focusing on monocotyledons and *Dicrostonyx* on dicotyledons; thus distinct food selection does occur in microtines. Kalela and Koponen (1971) found that the moss genus *Dicranum* was a favored food of lemmings in an area of isolated fjells in Finland, and they found correlations among decreasing altitude, decreasing percentage cover of *Dicranum*, and decreasing lemming density, although the omission of such variables as amount of protective cover makes these correlations less than definitive.

Extensive forage utilization seems to precede declines of brown lemming populations at Barrow, Alaska (Thompson, 1955a; Pitelka, 1973), but several other studies have evidenced neither obvious habitat destruction nor apparent starvation associated with declines (Chitty, 1960; Kalela et al., 1961; Krebs, 1964a). Although it is important to remember that forage limitations are not always conspicuous (Grange, 1965), this implies that while lemmings may sometimes feed copiously to compensate for some nutrient deficiency (Grant, 1978), this is probably a contributing but not a determining factor in lemming cycles.

Lemmings do, however, strongly influence the production of their habitat. Because the rate of vegetation decay is extremely slow in tundra regions, "the periodic conversion of the vegetation into lemming flesh and droppings is probably essential to the healthy growth of the plants on which lemmings feed" (Marsden, 1964). Also, the qualitative composition of plant material growing in the early spring will depend on nutrients stored and buds formed in the previous summer (Kalela, 1962), thus introducing a time lag into the population. Schultz (1964; see also Krebs and Myers, 1974) recorded grass production in a tundra site from which lemmings had been excluded and, over 7 years of observation (1958–65), noted that vegetation production in the exclosures declined while in grazed areas the production almost always increased (fig. 69). Since the

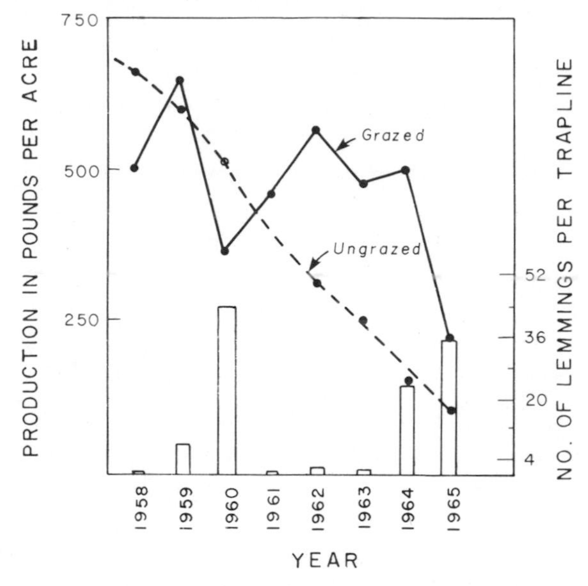

Figure 69. Primary production in exclosures (dashed line) decreases as a result of no lemming grazing. Primary production in open tundra fluctuates and is moderately related (correlation coefficient $r = 0.69$) to lemming density (bars). Production data from Schultz (1964); density data represent maximum average number of lemmings caught per trapline in the two or three trapping periods per summer from Pitelka (1972). (From Krebs and Myers, 1974. Reprinted with permission from *Advances in Ecological Research*. Copyright by Academic Press, Inc. (London) Ltd.)

exclosures were originally constructed in 1949, this may not be a simple relationship (Krebs and Myers, 1974). Dennis and Johnson (1970), in a study at Barrow, Alaska, found that lemmings can influence plant community composition by reducing litter and augmenting the depth of the active layer, and studies in the USSR demonstrate that grazing by lemmings can stimulate shoot production per unit area (see Bliss et al., 1973; Marsden, 1964). In Schultz's (1965) study, while increased lemming density seemed to decrease the vegetation cover, no significant relationship between the standing crop of vegetation and lemming density was observed.

Nutrient Cycling

Although there may not always be a connection between high lemming density and crop decimation, lemmings are a primary stimulant for nutrient cyles on the tundra. Pieper (1964), working at Barrow, found little variation in forage nutrient levels in areas where lemmings were excluded and found, like Schultz, that prevention of grazing lowered the nutrient content of the plants. Where lemmings were grazing, however, levels of phosphorus, nitrogen, calcium, potassium, and sodium (but not magnesium) were found at their highest concentrations in the forage during the summer of peak lemming density; Aumann (1965) independently recorded a similar observation for sodium. Pieper (1964) observed that the rapid winter population increase preceding summer population peaks would produce quantities of excretory products, which, with the spring melt, would supply large quantities of nutrients to the plants. This probably explains the synchrony between lemming density and nutrient levels (fig. 70).

The important question is whether food quality stops the increase phase of lemming populations. Hutchinson and Deevey (1949) considered the possibility that some limiting nutrient, which might be essential for animals yet relatively insignificant for plants, could precipitate the oscillation—they suggested sodium, based on observations that in some years hares suddenly die while on the run. Several of Collett's (1911) observations on migrating lemmings are similar to this observation in snowshoe hares: (1) migrating lemmings have a propensity to die of slight injuries; (2) "sometimes individuals will suddenly fall, have convulsions in their hind legs and under convulsions they will die in a few minutes"; and (3) "people say that they eat themselves to death on new grass." In fact, there is some evidence that sodium and phosphorous deficiency may occur in some vole populations (Grant, 1978). Alternatively, a salt

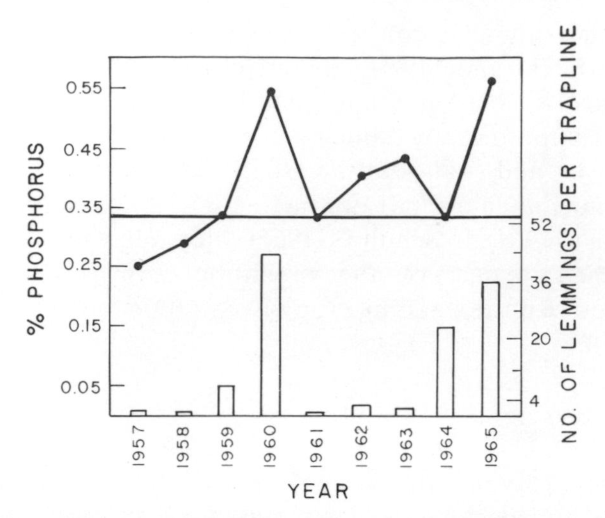

Figure 70. Percentage phosphorus in the summer forage and brown lemming densities (bars) in Barrow, Alaska. Horizontal line indicates similarity between phosphorus levels preceding and following years of peak lemming density. Data for phosphorus levels from Schultz (1969). Density data (Pitelka, 1972) represent maximum average number of lemmings caught per trapline in the one or two trapping periods per summer. (Redrawn from Krebs and Myers, 1974. Reprinted with permission from *Advances in Ecological Research*. Copyright by Academic Press, Inc. (London) Ltd.)

imbalance could be generated by prolonged adrenal hypersecretion under stress, which appears, for example, to cause mass mortality in sika deer, *Cervus nippon* (Christian, 1975). For microtine rodents Schultz (1964, 1969; also Pitelka, 1964) developed this idea into the concept of a nutrient threshold: forage nutrients peak during the summer of peak lemming populations, but high consumption at that time binds nutrients in organic material so that they will not be available to plants the next year. Nutrient levels in the forage might then be inadequate for, say, high reproductive levels, but the nutrients would gradually be restored over the next few years until forage nutrients and lemming densities recovered. No large lemming population increase could occur until a nutrient threshold could be reached. Schultz (1969) in this context particularly stressed the importance of phosphorus and calcium for reproduction.

Krebs and Myers (1974), relating the proportion of pregnant females to the proportion of phosphorus (important in reproductive processes and bone construction) in the vegetation, found no connection between phosphorous deficiency and lemming reproduction (it is not clear, however, what the more appropriate relationship between density of pregnant females with respect to available forage and proportion of phosphorus in vegetation might be). Krebs and Myers also noted that phosphorus levels were the same before and after the lemming peak and criticized the suggestion that phosphorus is a decisive factor, since those same levels of phosphorus giving rise to population increase are then assumed to be limiting. It is possible, on the other hand, that increased consumption could lead to an increase in total phosphorus intake when forage supplies are in excess of requirements. Another important point is that the breeding season in peak years is shortened, so that, for example, when nutrient levels were high in August 1960, the proportion of pregnant females was low (Krebs and Myers, 1974). Similarly, there is no clear indication that calcium limits lactation (as measured by mammary gland development; Mullen, 1968) or that sodium is a limiting factor (Aumann, 1965; Krebs and Myers, 1974). Thus the correlation between plant nutrient peaks and lemming population peaks may be more an example of the lemmings affecting the plants than of the plants determining lemming population densities. Observations at Barrow (Pitelka, 1973), where thriving vegetation for several years did not produce a peak in the local population but did provide forage for a large immigrant population, also suggest that the nutrient-recovery hypothesis may be inadequate to explain lemming cycles, but this situation is equivocal because this false population "peak" was not sustained through to the next summer, indicating that what appeared to be thriving forage might in fact have been somehow lacking in nutritive value.

It is conceivable that the search for a single nutrient determining the course of population cycles might be too narrow a viewpoint. Braestrup (1940) noted that "deficiencies affecting the acid–base balance are sufficiently widespread to account for the sudden drop in health observed in herbivorous mammals and birds at certain times." He referred to cattle licking disease and suggested that it was caused by a deficit in potassium and sodium in proportion to sulfuric and hydrochloric acids formed when grasses are digested. This proportion was found to be enormously variable from year to year depending on soil bacterial processes, which were in turn said to be influenced by the climate. A phenomenon of this nature might produce an indirect effect of climate or weather on cycle periods, but to my knowledge this question has not been pursued in studies of small mammal cycles.

Chemical Cues from Plants

Before becoming too skeptical of the role of herbivore–forage relationships as an explanation for microtine cycles, a wider range of subtle relationships should be considered. Freeland (1974) hypothesized that the loss of individuals from vole populations at peak density could be the result of reduced viability induced by toxic compounds in plants or plant parts consumed. Small amounts of plant toxins consumed by rodents can have many adverse effects, including inhibition of protein synthesis, inhibition of growth and reproduction, and other phenomena often recorded in cyclic species. Freeland noted that several qualifications would need to be met for the hypothesis to be valid: (1) preference for nontoxic foods as the result of susceptibility to the physiological effects of plant secondary compounds; (2) reduced availability of nontoxic foods with increasing population density, accompanied by increased availability of nonpreferred toxic foods; (3) increased ingestion of toxic foods at high density, leading to observed growth and reproduction decline and other effects; and (4) reduced cropping with population decline with preferred foods outcompeting toxic foods to set the stage for another population increase. Evidence from cycling *Microtus californicus* was analyzed in support of this idea.

One difficulty with this hypothesis, which Freeland noted, is that we have no clear idea of what is toxic (Batzli and Pitelka, 1975). If food preferences are tested based on (1) frequency of food appearing in the stomach, (2) quantity of food appearing in the stomach, or (3) an index comparing feeding behavior to the natural occurrence of plants in the field, no statistically significant correlation has been found (Schlesinger, 1976). It also seems doubtful that food preference is based solely on toxic compounds, without consideration for nutrient quality (Schlesinger, 1976). Besides, there is no clear evidence that decreased relative availability of preferred foods results in increased consumption of toxic plants: although preferred foods are in low supply during high densities of voles (*M. californicus*), there is no significant increase in intake of "toxic" plants at higher population densities. Also, seasonal shifts in food habits are apparent, but no consistent pattern in the proportion of the diet contributed by vegetative parts of different plants at different points in the cycle is evident; and the records for the same species at two sites with quite different vegetation both appear to cycle (Batzli and Pitelka, 1975). Thus the role of toxic secondary compounds in plants, although conceptually interesting, seems as yet unproven.

This does not mean to suggest that secondary plant compounds might not play a significant part in the cycle pattern. Berger et al. (1977) found that two naturally occurring cinnamic acids, and their related vinyl phenols (generated during the heat extraction of the cinnamic acids), inhibited reproductive function in *Microtus montanus*, as evidenced by decreased uterine weight, inhibited development of small ovarian follicles, and reduced breeding performance. Analysis of salt grass, comprising over 90 percent of the *M. montanus* diet in the Utah area examined, revealed that the cinnamic acids are most abundant in the food source at the end of the vegetative growing season, which coincides with the end of the breeding season for *M. montanus* in this habitat. When a group from a nonbreeding winter population was given limited supplements of fresh green wheat grass over a 2-week period, all of the experimental females became pregnant (Negus and Berger, 1977). What this implies is that the voles may cue reproductive effort from chemical signals in the plant resources, receiving a stimulative cue from a compound in actively growing plant tissue; inhibitor compounds increase in concentration as the grasses approach flowering and fruiting and, with the proposed simultaneous disappearance of the as-yet-unidentified stimulator compounds, reproduction ceases (Negus and Berger, 1977). Cuing reproductive effort to maximum food availability provides a useful adaptation to herbivores living in unpredictable environments.

Could this be related to microtine cycles farther north? This area is as yet unelucidated. One foreseeable difficulty is that, just before the population peak, lemmings reproduce excessively under the snow, quite past the end of the growing season, and then, in the peak summer, reproduction usually ceases abruptly in June (Krebs and Myers, 1974), at the peak of the plant growing season. Conceivably, the presence of a stimulatory compound in the annual summer growth would precipitate population increase until that growth was no longer available, but this implies a decimation of forage that we have seen is not always observed. If a stimulus, or inhibitor, is cycling through the forage, we are returned to the already observed difficulties of explaining longer-term cyclic regularity. Perhaps further investigation will reveal compounds providing clearly reliable cues to indicate the termination of a quality food supply. Until then, this remains a most attractive hypothesis.

A more convincing possibility for the effect of the termination of a quality food supply is suggested by recent observations of snowshoe hare (Keith and Windberg, 1978; in Windberg and Keith, 1978). Early regression of hare testis weight appears to be closely linked to food shortage the previous winter. This early regression indicates termination of the hare breeding season before the fourth, and last, litter could be conceived. Other studies have been unable to demonstrate adverse effects of high population densities on reproduction, suggesting that insufficient food is the determining factor. It is likely that lemmings, because of their exposed tundra habitat and susceptibility to occasional heavy predation, cannot surface to forage selectively and compensate for this during periods of nutrient insufficiency by copious consumption, essentially mowing the available forage in prepeak winters (Grant, 1978). This mass consumption in prepeak winters and the early termination of lemming breeding in peak summers, both suggest the possibility that nutrient shortages are involved in their population decline. It seems reasonable that nutrient deficiency could precipitate the population decline, or perhaps temper the increase. Predation might then force the population to a nadir; then a period of nutrient recovery would avert an immediate postpredation population boom. What remains to be explained is why, considering the effect of severe climatic variations, particularly revealed by the recent work at Barrow, the cycle periods of lemmings tend to cluster around 4 years.

Snowshoe Hare–Vegetation Interactions

Grange (1949) appears to have been the first to view the hare cycle as a hare–vegetation interaction. He suggested that postfire successional forest vegetation provided ideal forage for rapid hare population growth to a peak; overbrowsing and continued growth of browse beyond the hares' reach would produce a food shortage and subsequent decline. Grange (1965) also emphasized that it might not be easy to recognize when the winter food supply was depleted by hare browsing since the hares' winter diet is composed largely of recent growth of small twigs and stems. Fox (1978) suggested the existence of a cyclic flux in areas burned by forest fires in support of Grange's thesis and provided interesting evidence in support of this thesis. However, the hare cycle is basically a local phenomenon, and province-wide or nationwide cycle in forest fires would not explain a local fluctuation, even if immigration were invoked to some degree: the fact that some large area is destroyed by fire each decade would not necessarily explain a cycle in a distant area, although local synchrony might result. If fires do evidence cycles of a period similar to the wildlife, this could be connected to a common influence that both makes the forest particularly susceptible to fires, and diminishes the availability of hare browse; drought comes immediately to mind. However, little success has been attained in the search for a cyclic weather variable of the appropriate frequency.

Keith (1974) has summarized effectively (italics mine): "I believe that the basic hare–vegetation interaction occurs within all acceptable habitats, and that these may remain acceptable and indeed even favorable for hares over *several* cycles without intervening fire. I regard fire as extremely important in maintaining large areas of boreal forest in successional stages suitable for hares, but *not as a primary cause of cyclic fluctuations* . . . the occurrence of highest peak populations in the most favorable (fire produced) habitats does not necessarily mean that the cyclic fluctuations per se are dependent upon them. It simply means that population growth rates are greatest there, probably as a result of higher survival and reproduction, and lower rates of egress of both native and ingressing young."

Recent evidence strongly supports the importance of a herbivore–forage interaction in hare cycles. Wolff (1978b), working near Fairbanks, Alaska, examined browsing intensity and diameter at point of browsing (dpb, measured from randomly selected hardwood shrubs in the spring of each year) for snowshoe hare populations at a peak and subsequent low. Several facts emerged: (1) woody browse and spruce needles form a large part (about 82 percent) of overwinter food supply and are less important in the summer (fig. 27); (2) the dpb for a population high at one site in 1972 was significantly more than when the population decreased (fig. 71); (3) when the hare population decreased below its food-carrying capacity (in 1974), hares were able to forage on small twigs (less than 3 mm in diameter); (4) this diameter size remained relatively constant even though

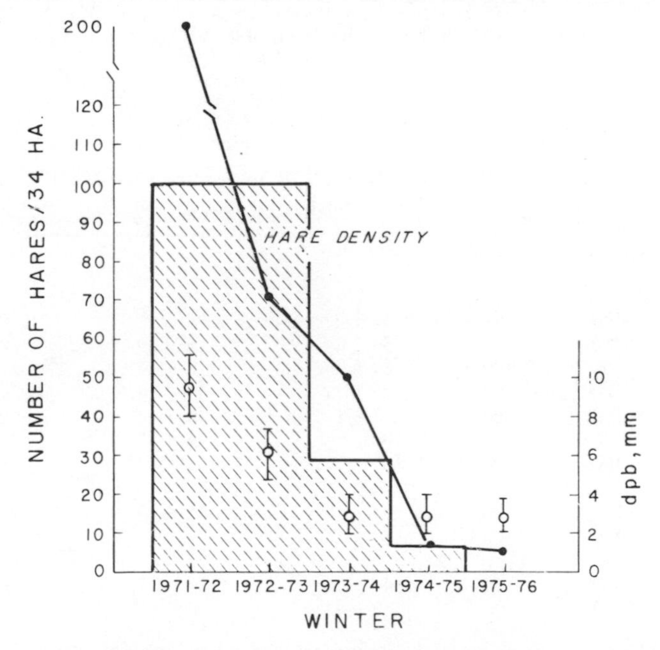

Figure 71. Snowshoe hare densities and diameter at point of browsing (dpb) for Wickersham control area, 1971–76. For dpb, Mean and range $N = 50$. Shaded area represents browse intensity. (After Wolff, 1977b.)

TABLE 3 Nutrient Analysis of Sixteen Basic Elements in 3-, 6-, and 10-mm-Diameter Willow and Alder Twigs from the Burn and Control at Wickersham.

	Zn	*Cu*	*Ca*	*Mg*	*K*	*N*	*Al*	*Fe*	*Mn*	*Pb*	*Co, Cr, Ni, Se, Hg, Mo*
Willow											
Control 3 mm	125	23	3890	1100	5040	123,760	142	68	8	1	All levels are less than 0.5
Control 6 mm	13	11	1560	835	2630	58,240	153	65	7	1	
Control 10 mm	23	11	1310	705	2060	60,480	149	70	6	1	
Control bark	45	12	6770	1060	5500	126,280	214	75	8	1	
Control wood	13	1	930	730	1540	40,320	109	60	5	1	
Burn 3 mm	127	9	5690	1155	4140	103,080	131	66	9	1	
Burn 6 mm	47	10	2070	850	2380	54,600	156	69	7	1	
Burn 10 mm	27	10	1600	735	2100	48,720	160	68	5	1	
Burn bark	119	11	5800	1260	5590	125,160	227	73	9	2	
Burn wood	15	8	770	715	1080	24,920	121	62	6	1	
Alder											
Control 3 mm	6	12	2360	1200	4410	120,400	126	74	5	1	
Control 6 mm	8	11	1360	950	2920	78,120	119	72	5	1	
Control 10 mm	21	6	1420	810	2430	62,720	130	69	7	6	
Control bark	14	11	3350	1085	4120	110,880	166	74	9	1	
Control wood	31	1	935	795	1730	37,520	102	61	6	3	
Burn 3 mm	6	8	3320	1285	4590	144,480	131	75	9	1	
Burn 6 mm	34	9	1740	955	2500	94,920	136	75	7	1	
Burn 10 mm	3	11	1460	865	2260	87,920	125	73	8	3	
Burn bark	12	9	4000	920	4030	156,800	137	76	13	1	
Burn wood	29	2	680	935	1810	78,680	113	63	7	3	

Source: From Wolff (1977a).
Note: Samples were collected in April 1975 during the third year of the hare decline. Concentrations in parts per million.

TABLE 4 Mean Browse Intensities by Snowshoe Hares at All Study Sites, 1971–1976

	Browse Intensity (%)				
	1971–72	1972–73	1973–74	1974–75	1975–76
Wickersham					
Area I	100	91 ± 2.4	10 ± 2.4	0	0
Area II	100	100	34 ± 2.6	4 ± 1.9	0
Area III	100	100	89 ± 4.4	3 ± 2.1	3 ± 2.9
Area IV	100	100	45 ± 3.6	4 ± 2.9	2 ± 2.0
Burn	100	100	3 ± 1.7	0	0
Goldstream I	100	100	100	99 ± 1.0	50 ± 4.0
Goldstream II	100	100	100	55 ± 4.3	25 ± 3.8
Bonanza Creek					
Refuge	100	100	99 ± 1.4	85 ± 3.7	21 ± 4.0
Open	100	100	40 ± 3.0	0	0
Tanana River					
Refuge	100	100	40 ± 2.7	20 ± 2.7	9 ± 1.0
Open	100	92 ± 4.1	69 ± 4.6	0	0

Source: From Wolff (1977a).
Note: Mean ± 1 standard error.

the population continued to decline for several years; (5) smaller twigs had quantitatively higher concentrations of 5 of 16 elements analyzed (Zn, Ca, Mg, K, N) than did large twigs (table 3); nitrogen, a measure of protein, was nearly twice as concentrated in 3-mm twigs as 6- or 10-mm twigs; (6) most of the nutrients were in the bark; and (7) browsing intensity was greatest during the population high, and declined to a low as the population decreased (table 4).

The lack of food as a cause of decline in the 10-year cycle was suggested by MacLulich (1937), Grange (1949), and Bider (1961). Keith (1963) proposed that during peak years the hare population exceeded its food-carrying capacity, resulting in malnutrition and starvation; Keith retained this idea in his recent model (1974), and quantitative evidence from studies near Rochester, Alberta, supports this idea (Windberg and Keith, 1976a). The main herbivore–forage interaction appears to occur between the hare population and its overwinter food supply of woody browse. The necessary component of this browse consists of stems less than 3 mm in diameter, mostly new growth from the previous summer, and hares are not able to maintain body weight on coarser stems even with unlimited browse (Pease and Keith, 1973, unpublished; in Wolff, 1977a; Pease et al., 1979). High browse intensity on coarse stems indicates food stress, since hares feeding on larger (6 to 15 mm) stems will need to consume considerably more browse and spend more time foraging than if they were feeding on smaller stems, since the nutrient content of stems decreases with increasing diameter. Although nutrient requirements of snowshoe hares are unknown, undernutrition is directly correlated with consumption of larger stems (Wolff, 1977a).

Girdling of trees and larger stems, infrequent during low or moderate hare densities, can be considerable during peak years (MacLulich, 1937; Grange, 1949; Bider, 1961; Windberg and Keith, 1976a; Wolff, 1977a). A moderate browsing intensity removing current annual growth can actually increase forage production the following growing season by having a pruning effect (Wolff, 1978a), but high-intensity browsing may reduce all new growth for several years, lowering the carrying capacity for hare; lateral branching of browsed terminal shoots, however, results in increased production of aboveground biomass over the long term (Wolff, 1977a).

This evidence supports a possibly intense, if occasionally equivocal, interaction between herbivores and their forage. Could this in itself generate the regular periodicity observed?

V. Volterra

When late in 1925 Umberto D'Ancona asked his father-in-law-to-be, the eminent mathematician Vito Volterra, how he might explain the fact that during and after the war, when fishing was restricted, the proportions of species of fishes caught in the upper Adriatic had changed significantly, the result was the formulation of what continues to be the major part of the deterministic theory of population dynamics. When viewed in combination with the work of Lotka, Kostitzin, and Kolmogoroff, one could say that essentially nothing new has been added to the theories of deterministic populations since that time (Rescigno and Richardson, 1973). What makes Volterra's work especially useful, however, is not just the application of elementary techniques leading to powerful solutions, but the emphasis on general principle, the effort to describe the eventual state of an ecological association. Beginning with intuitive ideas of how a population as a whole might behave, and then looking at interaction of different populations, Volterra generated a view of macroscopic population phenomena that helped many scientists begin to search for ways to look at biological associations. To appreciate the scale of Volterra's achievement, one might consider Scudo (1971), Goel et al. (1971), and Rescigno and Richardson (1973).

It may be the search for an answer exceeding the present level of our ecological understanding that made the Volterra analysis, originally considered basic and essential, eventually unfashionable. As noted by Goel et al. (1971), this analysis seemed to fall from favor, not because any calculations based on the model did not fit observed data, but because certain features of ecological systems were omitted or oversimplified. "If the same kind of criticism had discouraged people from the investigation of models used in many-body physics, physicists would have never developed the intuition necessary for the understanding of the behavior of complicated real materials" (Goel et al., 1971, p. 232).

The Volterra equations appear to be the first formulation of the concept of density dependence in ecology (Lack, 1966), and it is interesting that the idea was introduced on the basis of a purely mathematical formulation. As a result, this concept was initially overlooked by ecologists, much as Mendel's work had been, until a logical basis was provided by Nicholson (1933), who explained that if a population persists within a restricted range, it must be controlled by factors that tend to cause decreases at higher densities and increases at lower densities. Despite this demonstration of logical necessity, the dearth of evidence for density-dependent mortality precipitated a controversy over the value of the concept (e.g., see Lack's review, Lack, 1966); judging from current work (e.g., May, 1976a), the controversy may have subsided.

Because the Volterra analysis has played a large and controversial role in the study of population cycles, it seems useful to examine it in detail. Volterra (1926) described the interaction of two species through a pair of equations

$$\frac{dN_1}{dt} = \alpha_1 N_1 - \gamma_1 N_1 N_2$$

$$\frac{dN_2}{dt} = -\alpha_2 N_2 + \gamma_2 N_1 N_2$$

When the two species, 1 and 2, inhabit a common territory, the number of encounters between the predator (2) and prey (1) is proportional to the product of the population sizes ($N_1 N_2$); the fraction of encounters (γ_1) resulting in the death of a prey decreases the number of prey and contributes to the increase of the predator population (γ_2). If the prey were in isolation, it would increase in proportion to the existing number (N_1) and its coefficient of autoincrease (α_1); the predator in isolation, lacking food, would decrease in proportion to existing number (N_2) and death rate (α_2). Lotka independently investigated these same equations for the theory of two-species competition (Lotka, 1920; 1925).

It is historically interesting to note that the mathematical paradigm of viewing interactions between two groups and/or processes as a pair of simultaneous linear differential equations emerged in several disciplines at nearly the same time. Sir Ronald Ross, in his pioneering studies on malaria (Ross, 1908; 1911), employed this mathematical viewpoint to describe the propagation of malaria (see Lotka, 1923, for a thorough analysis of these equations). Lotka (1910) used the same equations later applied to two-species encounters as a viable description of autocatalytic chemical reactions. Lanchester (1916; discussed in Morse and Kimball, 1951), a noted aerodynamicist, used this paradigm to describe combat between opposing forces during World War I, and in 1919, L. F. Richardson, who became famous for his meteorological studies of turbulence, constructed a similar theory of combat, apparently independently of Lanchester (see Richardson, 1960). It seems reasonable from the relatedness of these fields that all of these views derive from basic approaches in physics, and the question of which came first remains a moot point (see Thrall, 1978; special thanks to G. E. Hutchinson for noting these authors on combat theory).

Integrating these equations, Volterra (1927, discussed in Scudo, 1971) discovered something that is particularly important in this study: the fluctuations of predator and prey population sizes are periodic, and this period depends only on the coefficients of prey increase (α_1) and predator death (α_2), the coefficients of interaction (γ_1, γ_2), and an integration constant dependent on the initial population size. This result is, unfortunately, disturbing, since it is difficult to accept the widespread occurrence of cycles in various animal populations and different species with similar periods as being dependent on a perhaps accidental combination of five constants, different (theoretically) for each species and dependent on the initial population size. Volterra (1927) did develop a perturbation method to demonstrate that when the populations vary only slightly from the mean values, the resulting period of small oscillations is, to a first approximation, independent of the initial conditions and the coefficients of interaction; but the data in this study suggest that the concept of small variations near equilibrium is inapplicable here, so this particular form of the Volterra equations is not appropriate for these cycles.

Another important point is that the deterministic Volterra model has no mechanism for regression of fluctuations, so that the system is in a permanently marginal condition. If the prey–predator relationship is described as a stochastic birth-and-death process following Volterra mechanics, the results are quite different (Walker et al., 1976): the sustained oscillatory behavior predicted by the deterministic model lasts for only one-fourth cycle and then rapidly degenerates. The stochastic description requires that the variables be discrete (i.e., the death of one prey corresponds to the birth of one predator), so the analogy is limited, but this does emphasize that the most interesting feature of the Volterra system, the oscillatory behavior, is dependent on the way the system is described, and that description may be in itself rather limited.

Volterra (1937; Scudo, 1971) generalized his ideas for an association of n species:

$$\frac{dN_r}{dt} = \left(\varepsilon_r + \frac{1}{\beta_r}\sum_{s=1}^{n} \gamma_{rs} N_s\right) N_r \qquad (r = 1, 2, \ldots, n)$$

where $\varepsilon_r N_r$ represents the propagation of species r if isolated in its environment (ε_r = natural birth-minus-death rate; conventionally, this is often called r_m, the intrinsic rate of increase). The β's are positive quantities such that in a binary encounter the ratio of the number of s's lost (or gained) per unit time to the number of r's gained (or lost) per unit time is β_r/β_s. These β's can be interpreted as nutritional values to predators or weights of individuals. The term γ_{rs} represents the coefficient of interaction of species r with species s. All $\gamma_{rr} = 0$ and $\gamma_{rs} = -\gamma_{sr}$ for $r \neq s$, since it is assumed that all encounters are one-sided such that in an $r-s$ encounter, a gain for r is a loss for s; the sign of γ_{rs} determines whether an encounter is advantageous or fatal to species r. This crucial simplifying assumption means that the probability of capturing a prey at an encounter and the probability of being devoured during an encounter are the same. Also, each individual of species s consumed is transformed into an individual of species r. Since this assumption bears no relationship to biological reality, the extrapolated results should be treated as interesting, but questionable.

A stationary state, in which all population numbers N_r have steady-state values q_r, will have a nontrivial equilibrium if all the roots of the system of linear equations

$$\varepsilon_r\beta_r + \sum_s \gamma_{sr} q_s = 0$$

are positive. This is possible only if the number of species is even and not all ε_r have the same sign. Defining

$$v_r = \log \frac{N_r}{q_r} \qquad [N_r = q_r \exp(v_r)]$$

we note that as $N_r \to q$, $v_r \to 0$, so that v_r is a measure of deviation from equilibrium (Goel et al., 1971). Rewriting Volterra's n-species equations, we have

$$\frac{\beta_r}{N_r}\frac{dN_r}{dt} = \varepsilon_r\beta_r + \sum_s \gamma_{sr} N_s$$

Introducing $\varepsilon_r\beta_r$ from the steady-state equation

$$\frac{\beta_r}{N_r}\frac{dN_r}{dt} = \sum_s \gamma_{sr} q_s + \sum_s \gamma_{sr} N_s$$

bringing in the new dependent variables, and rearranging yields

$$\beta_r \frac{dv_r}{dt} = \sum_s \gamma_{sr}(N_s - q_s) = \sum_s \gamma_{sr} q_s (e^{v_s} - 1)$$

Multiplying throughout by $q_r(e^{v_r} - 1)$ and summing over all r (remembering that $\gamma_{rs} = -\gamma_{sr}$), we obtain

$$\sum_r \beta_r q_r \frac{dv_r}{dt}(e^{v_r} - 1) = 0$$

and integrating gives us

$$G = \sum_r \beta_r q_r (e^{v_r} - v_r) = \text{constant}$$

This means that, for an even number of species, there exists a universal single-valued constant of the motion, G, that is a sum of terms relating to the separate species in association. Thus whatever the initial value of G_0 for a system, that value G_0 will be maintained indefinitely. Generally, the more disparity between initial values of N_r and their equilibrium values q_r, the larger the value of G (Maynard-Smith, 1974).

Volterra's encounter method for population interaction derives directly from classical statistical mechanics and is comparable to the "law of mass action" in chemistry. This type of approach is especially useful when the details of species interactions are missing. In addition, the fact that G is a sum of terms each of which relates to an individual species allows a means of specifying components of this larger system in the sense used in statistical mechanics (Goel et al., 1971). Kerner (1957, 1959, 1961, 1964) used these ideas to construct a statistical mechanics for the Volterra system. The dynamical equation for the n-species Volterra system was used by Kerner to define the controlling factors for each system, and a Gibbs ensemble of Volterra systems was constructed such that each system represents a possible set of initial values $\{v_r\}$ consistent with the constant G. A point in phase space $v_1 \cdots v_n$ can describe the state of each member of the ensemble, and a collection of phase points represents the state of the ensemble. This collection of points moves through the phase space as the total ensemble evolves through time. It can be shown that the density of phase points in the phase space is conserved (Goel et al., 1971).

Suppose that the only knowledge one has about a biological association is that it has a constant G. Then define a canonical ensemble

$$\rho = e^{\psi} - G/\Theta$$

where ρ gives the probability of finding the system in the configuration $(v_r, v_r + dv_r; v_s, v_s + dv_s)$ when the system has been running a long time and has attained equilibrium; in this case Θ and ψ are analogous to temperature and free energy in physical systems. Moreover, assume that biological associations are ergodic; that is, time averages over a single system are equal to averages over a suitable ensemble from that system. This is comparable to saying that a system can forget its origins and past; for example, we can start with any initial age structure, apply some operation repeatedly, and the result will be the same age structure! Although this is generally not true, it is true in the linear approximation if the number of species is large (Goel et al., 1971). This provides a means for calculating the time averages, which are of primary concern here.

For the Volterra equations, Kerner (1959) found

$$\frac{\Theta}{\beta_r q_r} = \frac{1}{x_r} = \frac{\overline{(N_r - \bar{N}_r)^2}}{\bar{N}_r^2} = \overline{\left(\frac{N_r}{q_r} - 1\right)\log\frac{N_r}{q_r}}$$

where a bar indicates a time average and $\bar{N}_r = q_r$. This means that the temperature Θ is a measure of the mean-square deviation of each species from its equilibrium population. Since this Θ applies to all species in the ensemble, all species fluctuate with "equal energies," analogous to the idea in mechanics that all particles in a system have the same average kinetic energy. The true value of these averages is that they are related to one time-independent parameter x_r, which is accessible through the population data.

Following Kerner's (1959) example, these averages were calculated for several populations, and the results are tabulated (table 5). Although one would hesitate to make definitive statements, the x values for foxes in Canada and central Norway are remarkably consistent. Northern and southern Norwegian foxes are somewhat equivocal, as are lynx and hare in Canada. It is important to remember that this comparison of theory and observation is more a test of the assumed probability distribution of

TABLE 5 Parameters Based on Kerner's Theory: $Q = N_r/\bar{N}_r$

	Moravian Missions (Elton, 1942)				*Hudson's Bay Company (Moran, 1949) (MacLulich, 1957)*			
	Arctic		*Colored*		*Lynx*		*Hare*	
		x		x		x		x
$\overline{(Q-l)^2}$	1.596	0.6	0.797	1.3	1.054	0.9	0.534	1.9
$\overline{(Q-l)\log Q}$	1.423	0.7	0.750	1.3	1.132	0.9	0.783	1.3
$\overline{\log Q}$	−0.818	0.7	−0.412	1.3	−0.652	0.9	−0.497	1.1
$\frac{x_r}{q_r} = \frac{\beta_r}{\Theta}$	$\frac{0.7}{2.2} = 0.32$		$\frac{1.3}{2.3} = 0.57$		$\frac{0.9}{2.9} = 0.31$			

	Norway Fox (Elton, 1942)					
	South		*Central*		*North*	
		x		x		x
$\overline{(Q-l)^2}$	0.051	19.6	0.224	4.5	0.876	1.1
$\overline{(Q-l)\log Q}$	0.056	17.9	0.200	5.0	1.044	1.0
$\overline{\log Q}$	−0.029	> 10	−0.100	5.0	−0.611	1.0
$\frac{x_r}{q_r} = \frac{\beta_r}{\Theta}$	$\frac{17.9}{3.8} = 4.71$		$\frac{5.0}{3.1} = 1.63$		$\frac{1.0}{2.8} = 0.36$	

the canonical ensemble

$$P(n_r)dn_r \simeq n_r^{x_r-1}e^{-x_r n_r dn_r} \qquad \left(n_r = \frac{N_r}{q_r}\right)$$

than of the Volterra mechanics (Kerner, 1959) and does not necessarily verify the proposed scheme of interactions (Leigh, 1965). The fact that the numbers generated are consistent does not necessarily enhance our understanding of causal factors for cycles.

Assuming that the Volterra prey–predator concept is applicable and that Kerner's x values are meaningful in these terms, it is interesting to speculate on the meaning of x_r. Since

$$x_r = \frac{\beta_r q_r}{\Theta}$$

x_r incorporates predator–prey equivalents (β_r), equilibrium population values (q_r), and a measure of the stability of the population (in terms of the thermodynamic temperature, Θ). Values of $x_r/q_r = \beta_r/\Theta$ can be computed (table 5). We can assume that the β_r, which might include such factors as nutritional value and proportion of population trapped, are comparable in given areas. Thus, for a given area, a larger value of β_r/Θ indicates a smaller value of Θ, hence more stability. Thus the stability for each animal would increase in the order

Canadian lynx < Labrador arctic fox < Labrador colored fox

for Canadian poplations (note that lynx, from the forest area, might not be quite comparable to tundra mammals) and for Norway

North Region < Central Region < South Region

Although the medians in these data sets are comparable, it is not particularly surprising to discover that colored foxes are more stable than arctic foxes in Labrador, since the colored fox has the option of migrating into the forests for alternative food sources when tundra food supplies are scant.

It is surprising that there is such a striking increase in stability in Norway populations as one moves from the desolate northern coastal regions, through the mountainous central region (perhaps an optimum habitat from the point of view of the lemming prey) to the comparatively lush southern region, where the diet for the fox may be more diverse.

Generally, the statistical thermodynamics approach does not clearly define whether we are discussing the number of species or the number of associations among species. The theory thus yields an intriguing framework for organizing ecological thought, but with the pieces missing. This seems in reverse order to a naturalist, who tends to build his framework from the pieces: it seems comparable to skipping perception and going straight to abstraction. We must be prepared, however, to consider the possibility that many of the functional responses observed in various species are a secondary consequence of numerical responses which, when viewed on a larger scale from a higher level, may be generated by a small number of constraints relevant to the individual species, the environment, or the system as a whole. Wherever the answer lies, it seems important that both the naturalist-ecologist and mathematician understand each other's viewpoints in order to avoid the types of philosophical nonquestions that can become long-term major controversies in ecological sciences.

Kerner's basic mode of analysis has been reviewed and extended by Leigh (1968). Beginning with Kerner's (1959) ensemble average

$$\overline{v_p \frac{dv_r}{dt}} = \frac{\gamma_{pr}}{\beta_p \beta_r} \Theta$$

and assuming ergodicity, Leigh demonstrated that, for time period $-T$ to T,

$$\overline{v_p \frac{dv_r}{dt}} = \lim_{T \to \infty} \frac{1}{2T} \int_{-T}^{T} v_p \frac{dv_r}{dt} dt$$

$$= \frac{d}{ds} \left[\lim_{T \to \infty} \frac{1}{2T} \int_{-T}^{T} v_p(t) v_r(t+s) dt \right] s = 0$$

The center portion of this equation (between the equality signs) will be recognized as the slope of the covariance function taken at lag $s = 0$. The conclusions to be drawn from this are twofold: (1) in a many-species system subject to Volterra mechanics, correlation coefficients (which are simply covariances divided by a constant, the standard deviation) provide a reasonable method of measuring the coefficients of interaction γ_{pr}; and (2) for a species in a stable environment (assumed in the Kerner analysis) limited only by food-web relationships (assumed in the Volterra mechanics) the correlation function should be smooth at lag $s = 0$, resembling the maximum of a parabolic curve (Leigh, 1968).

Now consider a logistically regulated species in a random environment. A linear approximation of this stochastic logistic will be

$$dx(t, a) = -aKx(t, a)dt + dB(t, a)$$

where the integral of $dB(t, a)$ over the time interval $t_1 < t < t_2$ is a random variable with mean 0, variance $\sigma^2(t_2 - t_1)$, and a Gaussian distribution; a is an index of the number of possible ways of realizing $x(t)$ and varies between 0 and 1; and K is the equilibrium value of the population, which is also the maximum value that the population can attain.

Leigh derives the covariance function for this system,

$$\lim_{T \to \infty} \frac{1}{2T} \int_{-T}^{T} x(t, a)x(t + s, a)dt = \frac{\sigma^2}{2ak} e^{-aKs}$$

for $s > 0$, and for $s < 0$ the result is

$$\frac{\sigma^2}{2ak} e^{-aK|s|}$$

This means that for a logistically regulated species subjected to a random input, the autocorrelation function has a sharp peak at $s = 0$, falling off exponentially at either side of the peak.

Although these results are theoretically interesting, it is difficult to see how they can be applied. In a 4-year-cycle species, the autocorrelation function would be expected to drop from 1 to 0 in the first time lag and, since the autocorrelation function is symmetric about zero lag, it would be difficult to discuss the "smoothness" of the curve at this point. Even in the 10-year-cycle group there is usually a drop to less than 0.5 at r, which might be an indication of extensive external noise in the system (Leigh, personal communication), but again smoothness would not easily be ascertained.

Leigh (1968) proposed a model incorporating both Volterra mechanics and environmental noise:

$$dx_i(t, a) = \sum_{s=1}^{r} \gamma_{is} \frac{\partial G}{\partial x_s} dt + dB(t, a)$$

By assuming that the environmental inputs of all species are independent and that the variance σ_i^2 of each input is equal to $-2\gamma_{is}\Theta$, the equilibrium distribution and the derivatives of the cross-correlation function at zero lag can be shown to be the same as for the conservative theory (even number of species using Volterra mechanics). If the environmental outputs of different species are not independent (suggesting population regulation by competition as well as predation), the analysis becomes exceedingly complex, and it is best to refer to the original paper for details (Leigh, 1968). Letting $R = [r_{ij}]$ be the covariance matrix for population sizes, defining matrix $M - [m_{ij}]$ where $m_{ij} = \gamma_{ij} p_j$, and using linear approximations, it can be shown that

$$\lim_{s \to 0+} \frac{dR}{ds} = RM^*_{ij}$$

where $M^*_{ij} = [m_{ij}]$. Letting $R_{ij}(0) = 0$ for $i \neq j$ yields

$$\lim_{s \to 0+} \frac{dr_{ij}(s)}{ds} = m_{ji} r_{ii} = q_j \gamma_{ji} r_{ii}$$

so that the derivatives as $s \to 0$ of the covariance are the same as for the Volterra case. Leigh notes that this result probably applies to a wide range of stochastic equations and that the derivatives of the correlations may only be dependent on the linear approximations to the processes being described.

Whereas the visual criteria for examining the autocorrelation functions are difficult to apply to the oscillatory autocorrelations observed here, these last results can be amenable to testing. Leigh (1968) did, in fact, this test for the two-species case using the same Hudson's Bay Company hare and lynx data described here, and discovered from the matrix analysis that a period of oscillation of 25 years would be expected, a figure hardly comparable to the observed 10-year period. But there are several problems here: (1) the data sets may not be comparable, since the lynx data are from the Mackenzie River district of western Canada while the hare data are from the lower Hudson Bay area of eastern Canada; (2) the demonstrated phase difference between these two areas would generate correlation values for lags 0 and 1 that are significantly different from the values that would be generated if the oscillations were in phase; (3) these matrices are based on autocorrelation and cross-correlation functions and their derivatives, and strong autocorrelations, which have been demonstrated, can produce highly exaggerated cross-correlations, distorting the results; (4) there is some suggestion that the theoretical equations are too sensitive to variations in the covariances to allow convincing conclusions based on estimated covariances where frequencies of this magnitude are concerned; and (5) other hare predators may need to be included in the description.

There continue to be several difficulties with trying to use the Volterra equations in this form. The basic equations describe a conservative system, and it is necessary that the system be conservative in order to use the methods of statistical mechanics. Whether this conservative system will exhibit the violent fluctuations we are trying to explain depends entirely on the initial conditions. What produces these initial conditions remains a mystery, as does the reason why so many species evidence oscillations of the same period length. If the concept of a Volterra-like predator–prey interaction is valid, one must answer such questions as: What common factors between muskrats and snowshoe hares might produce oscillations of the same period length? If the interaction were between carnivores and prey (which seems unlikely), we could note similarities between predator–prey communities: both are prey to foxes, hawks, owls, and other less specialized predators; both have a worst enemy, mink for the muskrat and lynx for the hare; both are herbivores feeding on a food base that is productive only at certain times of the year (Seton, 1909). But the concept of common predators suggests that the Volterra equations might need to be modified for prey switching, although if the two prey species have the same intrinsic growth rates (which may not be true; Tanner, 1975), a stable multispecies equilibrium may result which approaches a Volterra oscillation (Tansky, 1978). And, although mink and lynx may exert quite a bit of predation pressure on muskrats and hares, respectively, the mink is more omnivorous, and would therefore theoretically have means to escape these violent fluctuations. Finally, although both are herbivorous, the muskrat lives in a marshy habitat where food material is quickly (seasonally) regenerated, whereas the snowshoe hare occupies swamps, thick forests, and riverside thickets (Banfield, 1974), where vegetation recovery time after a hare peak might take on the order of several years (Keith, 1974). The latter facts also argue against a simple Volterra cycle generated by herbivore–vegetation interaction, unless a more subtle form of vegetation recovery, such as depletion of vital nutrients, is invoked.

Another ecological aspect neglected in the original equations (an omission that Volterra recognized) is the absence of a saturation term. In the absence of predation, the population of a species will saturate and not grow indefinitely (Verhulst, 1845; see Goel et al., 1971). If a Verhulst saturation term $dN_r/dt = \alpha_r N_r(\theta_r - N_r)/\theta_r$, where θ_r is the saturation level for species r, is included in the n-species Volterra equations, the fluctuations in population for a given species will be damped out; the time scale of this damping effect may be quite long, however, so a large number of oscillations may occur during the damping time (Goel et al., 1971; Rescigno and Richardson, 1973). The saturation term might vary with time, probably altering the system's dynamics. Another weakness of these equations is that prey density has no effect on the likelihood of a prey being eaten, and predation density has no effect on the likelihood of a predator catching a prey (Smith, 1952), which seems biologically unreasonable.

Yet another problem made Volterra skeptical of whether these equations might be good representations of natural systems: a stationary state can only exist for an even number of species. For example, there can be no equilibrium distribution among three species that will produce finite nonvanishing populations for all three species. However, including other factors, like predator switching, may allow for a stable equilibrium for an odd number of species.

One way to approach this problem is to include the saturation term in coefficients of increase ε_r by assuming that these are linearly decreasing functions of their numbers; that is, all γ_{rr} are negative (Scudo, 1971). Even one self-limiting species with negative γ_{rr} would be more biologically acceptable. Volterra (1926) suggested that one might consider a simple food web with an odd number of levels: for example, a carnivore preying on a herbivore grazing on a single plant species (Scudo, 1971). This can be described as:

$$\beta_1 \frac{dN_1}{dt} = (-\beta_1 \varepsilon_1 + \gamma_{12} N_2) N_1 \qquad \text{(carnivore)}$$

$$\beta_2 \frac{dN_2}{dt} = (-\beta_2 \varepsilon_2 = \gamma_{21} N_1 + \gamma_{23} N_3) N_2 \qquad \text{(herbivore)}$$

$$\beta_3 \frac{dN_3}{dt} = (\beta_3 \varepsilon_3 - \lambda N_3 - \gamma_{32} N_2) N_3 \qquad \text{(plant)}$$

where all constants are positive and where λN_3 represents the self-limitation for the space-limited plants (Scudo, 1971; Goel et al., 1971). It can be shown that the fate of the carnivore depends on whether or not

$$(\gamma_{12}\gamma_{23}\beta_3\varepsilon_3 - \gamma_{23}\beta_1\varepsilon_1 - \gamma_{12}\lambda\beta_2\varepsilon_2) > 0$$

Whether the equilibrium point is approached monotonically or through damped oscillations depends on the values of the parameters (Scudo, 1971). One must not assume that these oscillations will be periodic: periodicity from Volterra equations is only clearly demonstrated for the two-species, predator–prey case (Rescigno and Richardson, 1973).

Also, we need to know quite a bit more biology to know whether we have included a superfluous factor. An enlightening study of the effects of continuous time lags on a Volterra system, including a predator and two competing prey (Caswell, 1972), shows the wide range of behavior available for such a system, including the effects of responses to various types of competition, "hunger" of the predator population (leading to size selective predation), changes in prey abundance, food requirements of predators, and heterogeneity of the environment. While Caswell's study emphasizes the wealth of possible explanations available, it also obscures the nature of cycles. Only firmly established, clearly applicable biological fact can change possibilities into explanations.

Finally, the Volterra system is structurally unstable: a slight alteration in the mathematical formulation of the various terms can alter the dynamics completely (May, 1976b). It may be time to stop discussing systems such as the Volterra prey–predator equations as possible explanations for cycles, and to begin searching for something less restrictive.

Generalized Deterministic Equations

One particularly interesting extension of the two-dimensional predator–prey model, inspired by Volterra's work, was developed in a general form by Kolmogoroff (1936), and has been reviewed in detail by Rescigno and Richardson (1967). The generalized equations examined were:

$$\frac{dN_1}{dt} = N_1K_2(N_1, N_2) \qquad \text{(prey)}$$

and

$$\frac{dN_2}{dt} = N_2K_2(N_1, N_2) \qquad \text{(predator)}$$

where K_1 and K_2 are continuous functions of population size and have continuous first derivatives. The biological restrictions are:

1. $\partial K_2/\partial N_2 < 0$; for a given population size (numbers, biomass, etc.), the rate of increase of the prey species is a decreasing function of the number of predators.
2. $N_1(\partial K_1/\partial N_1) + N_2(\partial K_1/\partial N_2) < 0$; for a given ratio between predator and prey the rate of increase of the prey is a decreasing function of population size.
3. $K_1(0, 0) > 0$; when both populations are small, the rate of prey increase is positive.
4. $K_1(0, A) = 0$ for $A > 0$; there is a predator population size A which is large enough to block further prey increase, even when the prey is rare.
5. $K_1(B, 0) = 0$, for $B > 0$; there is a prey population size B which is an upper limit even in the absence of predators (resource or other limitations).
6. $\partial K_2/\partial N_2 < 0$; the rate of predator increase decreases with their population size.
7. $N_1(\partial K_2/\partial N_1) + N_2(\partial K_2/\partial N_2) > 0$; for a given ratio between the two species, the rate of predator increase is an increasing function of population size.
8. $K_2(C, 0) = 0$, for $C > 0$; there is a prey population size C that stops predator increase even if the predators are rare.

9. $B > C$; otherwise, the predators will disappear (Rescigno and Richardson, 1967; Scudo, 1971; May, 1973a).

Under these conditions, the described system will have either a stable equilibrium point or a stable limit cycle. Thus stable oscillatory behavior may be found in models more complex than quadratic models. Additionally, this model essentially describes most of the conventional models in the ecological literature (May, 1972, 1973a), including the conservative Volterra system, but without requiring that the amplitude and period of the observed oscillation be dependent wholly on the initial conditions.

Cycles in small mammals may well be stable limit cycles, perhaps with superimposed environmental fluctuations, with the limits being defined by interactions between and within species. However, it is useful to be cautious here. We are working with a large number of different species, especially in the 10-year cycle, all of which evidence cycles of approximately the same period, and with some suggestion of continent-wide phase variations. Not all predator species pursue the same prey, and some predators have wider food preferences than others and are thus not as dependent on a single prey species. It is difficult, therefore, to understand why so many species have the same period length unless predators are following prey cycles, as suggested by the fact that colored fox, *V. vulpes*, fluctuates with a 4-year period on the Labrador coast and a 10-year period in the boreal forest. If the prey species are the determining factor and are, for example, in a stable limit cycle with respect to their plant forage, it is not immediately obvious why the various parameters for snowshoe hares and muskrats should generate cycles of the same period unless hares are producing the cycle and muskrat cycles are generated by predator switching, in which case it would be difficult to explain why the muskrat population appears to peak about 3 years before the hare, or unless both are being cued environmentally, and an environmental cue that can produce continental phase variations does not seem likely. Although stable limit cycles may be a major component of the observed oscillations, other factors may require consideration.

It is interesting to note that, if we are dealing with stable limit cycles here, we may be observing phenomena suggested by Kolmogoroff's theory, namely that the population's dynamics may switch from a stable limit cycle to a stable equilibrium point if circumstances change. This may explain the marked fluctuations in foxes in central Norway, but not in the north or south of Norway (Finerty, 1972), a situation comparable to the larch budworm in Switzerland, which shows fairly regular oscillations in its optimal subalpine habitat, but significantly fewer regular changes in less favorable habitats above and below this area (Baltensweiler, 1971; May, 1972; Hutchinson, 1975). Another way to view this idea of optimal habitat is the "paradox of enrichment" (Rosenzweig, 1971). If the environmental carrying capacity for the prey is much larger than the equilibrium prey density when predators are present, the contribution to dynamic stability by the prey density dependence will be relatively weak, so that increasing the carrying capacity will lower the stability and eventually result in stable limit-cycle behavior. However, although this view may be appropriate for the larch budworm (May, 1976c), it may not be directly applicable to the foxes, where the carrying capacity, as measured by prey availability, is itself fluctuating.

Unfortunately, the Kolmogoroff model may not apply beyond two-species systems (May, 1973a). It would, therefore, seem useful to verify that the systems under study represent highly dependent two-species relationships, or that one can select a prey and group all predators without having to include questions of competition, which might invalidate the assumptions of the model. Clearly, there is more work to be done in this area.

Explorations of dynamic stability of prey–predator or herbivore–forage models (reviewed in May, 1976c; see also Tanner, 1975) suggest a tendency toward stable cycles if the intrinsic growth rate of the prey exceeds that of the predator and the carrying capacity of the environment is relatively large. The first criterion is probably met in the populations being considered here. However, the carrying capacity is highly variable, owing to essential denuding of forage both by microtines (e.g., lemmings; Pitelka, 1973) and snowshoe hares (Keith, 1974). For a plant–herbivore system, the growth rate of the prey (plant) relative to the predator (herbivore) may be quite low, owing to time lags in forage recovery (e.g., see Keith, 1974). It is possible that this view of stability may not prove to be useful in these cyclic systems.

Relationships between One- and Two-Species Models

Two-species models and single-species models can be directly related. The growth rate of the prey population will depend on the predator population, which can be represented as some integral over past prey populations. This means that an equation for the prey population can be developed involving only the prey population, where the density-dependent regulatory mechanism is expressed in terms of past values of the prey population. If the per capita numerical response function of the predator is independent of predator numbers, thus depending only on prey density; and if prey increase depends nonlinearly on both prey and predator population densities, the result will be an equation with a time-delayed regulatory mechanism similar to the delay-differential logistic previously discussed (May, 1976c), where the time delay represents the time for predator populations to respond to prey changes, or, for herbivores, the time for vegetational recovery.

McMurtrie (discussed in May, 1976c) offers a variation of this idea for a herbivore–plant or plant–substrate system, a variation applicable to any predator–prey system. In his model the "predator" population $N(t)$ has a logistic growth curve, and the carrying capacity $K(t)$ of the "prey" resource depends on $N(t)$ as

$$\frac{dK}{dt} = \frac{K_0 - K(t)}{T} - bN(t)$$

where K_0 is the resource saturation level and b the rate of depletion by consumers. Resource recovery rate depends inversely on T, the intrinsic resource regeneration time, and directly on $K_0 - K(t)$, the extent below saturation to which the resource is depressed. Integrating to find $K(t)$ and substituting this for the carrying capacity in the ordinary logistic, an equation of the form of a time-delayed logistic for the "predator" population results. Like the simple logistic, the equilibrium point

$$N = \frac{K_0}{1 + bT}$$

is stable, but unlike the simple logistic, when

$$b > \frac{(1 - rT)^2}{4rT^2}$$

perturbations die away in oscillatory fashion. Thus high depletion rates, which could be predators depleting prey, herbivores depleting plants, or plants depleting nutrients, all of which evidence high depletion rates in these cycles, tend to produce oscillations.

Similar to the simple time-delayed logistic, we find a solution with a stable equilibrium point giving way to stable limit cycles once the ratio of the time lag to the natural response time of the system exceeds a value of the order of unity (May, 1976c). Once the stable limit cycles arise in equations of the general form of the time-delayed logistic, their period, as we have seen, is roughly 4τ (May, 1976b). It has been noted that if we assume that the intrinsic growth rate r for the "predator" is constant (although we have seen that this is unlikely), and that r is large enough that the response time ($1/r$) is much less than τ, the predators would track a periodic flux in the carrying capacity, or, in this model, predators would track prey flux. The period of this flux would be roughly 4τ, so that, in the 4-year cycle, for example, τ would need to be near unity, or slightly less. Now what does τ represent in this instance? It is related to the rate of depletion by consumers, the intrinsic regeneration time of the resource, and the degree to which the resource is depressed below saturation, but its exact mathematical definition is not clear, so it is difficult to have any degree of certainty concerning the possibility that this particular formulation can account for regular periodicity. However, one possibility is reinforced: it seems more likely that some intrinsic mechanism is generating the cycles and that "predators," on whatever trophic level, are tracking fluctuations in prey rather than producing oscillations through species interactions. We have also seen that this idea is clearly supported by the biological facts.

Because single- and two-species population dynamics can be directly related, it is not surprising to find a dynamic complexity in simple predator–prey models as unexpected and disconcerting as that observed in the single-species case (Beddington et al., 1975, 1976). How valuable to recognize, after decades of study, that even the simplest mathematical models can generate complex fluctuations, including cycles. Since the experimental observations necessary to explicate which natural mechanisms are significant in the field are difficult to obtain, so that we cannot clearly know that our simplifying assumptions are not too restrictive, and since our recent mathematical discoveries reveal a new range of dynamic possibilities for previously suggested, and new, models, it may be necessary to admit, as Guckenheimer et al. (1977) suggest, that "given our present state of knowledge of population dynamics—both biological and mathematical—it is premature to make general assertions concerning the causes underlying fluctuations in real populations."

Extrinsic Factors: Weather and Celestial Influence

The facts that the period of population oscillations is greater than the average life span for most cyclic species, and that synchrony may occur over quite large areas, suggest that some factor external to the individual organism might be producing or inducing the oscillation. Weather cycles might be an important factor here.

Climate is a contributing factor limiting the range of most species. As numbers increase, competition may force the surplus into a part of the environment that is less stable, perhaps unstable, and weather changes could severely affect populations in such areas (Solomon, 1949; Moran, 1954). Weather may also influence population processes by acting as a source of perturbation for the system. Seton (1911) suggested this for the muskrat. Because the food and predators of muskrats did not appear to fluctuate in the dramatic way the muskrats seemed to, Seton attributed the muskrat decline to "a sudden great rise of the water after the ice has formed, so that the Rats are drowned; or to a dry season followed by severe frost, freezing most ponds to the bottom, so that the Rats are imprisoned to starve to death, or are forced out to cross the country in winter, and so are brought within the power of innumerable enemies." The regularity of the muskrat flux was attributed to the "well-known" cyclic variation in the amount of water in the northwest, a suggestion which is, to my knowledge, undocumented. Butler (1962) provided evidence that muskrat populations may be correlated with rainfall in areas where rainfall actually creates habitats that can later be eliminated in dry years (e.g., in southern Saskatchewan, where muskrats live in sloughs that can dry out during drought).

Weather perturbations have also been suggested for the lynx cycle. Fitting the Hudson's Bay Company Mackenzie River lynx data to a second-order autoregressive process, Moran (1953b) found a strong correlation between the residuals of this process and temperature (minimum temperature for the winter 2 years before trapping and summer temperatures for the summer before). He thought that these temperatures might influence the birth rate, perhaps by affecting the abundance of snowshoe hare (either directly, or indirectly through its vegetative food sources). Watt (1969), using the same second-order autoregressive fitting, found a strong correlation between the residuals and global temperature data, and concluded that the process is basically endogenous but subject to disruption from outside forces. But if the process is autoregressive of the second order, rapid damping of the autocorrelations would be expected, and the present study (of the same data) does not support this, so that this model may not be adequate. It is, however, conceivable that a random perturbation occasionally depresses the nadir of the cycle so that the amplitude of the oscillation is increased, thereby decreasing the effect of damping.

Krebs and Myers (1974) reviewed the evidence for the influence of weather on population cycles in small mammals. The evidence suggested that the critical period for weather effects for several microtines was in the autumn, with a low-survival period preceding harsh winter weather, and in the spring, associated with the initiation of breeding. Johnsen (see Elton, 1942) had suggested this possibility, connecting lemming migration years in Sør Trøndelag with an early spring following a late autumn the previous year, with an early winter often occurring in the year of the migration. Collett (1911) had noted that the mountains near Sør Trøndelag experienced the most frequent lemming migrations, but that these local migrations were not every 4 years, whereas the Norway fox data, used as an indicator of lemming abundance, suggest a local flux with a period of 4 years. This can be interpreted as implying that optimal weather circumstances can lead to a population peak extreme enough to precipitate a migration, but this does not of necessity imply a connection between weather and regular periods. Wildhagen (1952) drew the same conclusion when trying to connect lemming highs with summer temperatures: peaks may be less prominent in warm summers than in cool summers, but they still exist.

Shelford (1943) recorded that heavy snowfall, especially in the first 3 months of winter, favored lemming (*Dicrostonyx*) increase. Fuller et al. (1969) studied the tundra environment near Great Slave Lake, Northwest Territories, in detail and determined that several factors needed to be considered, because an area that appears uniform on the surface might evidence large variations in microclimates below the snow, where microtines spend the winter. After determining that subnivean temperatures were poorly correlated with air temperature, Fuller and his group recognized the need to look at variables such as snow thickness and rate of snow accumulation in addition to air temperature. They also noted that different genera responded differently to similar conditions, so that lack of synchrony between genera should not be interpreted as denying weather as a control factor. In conclusion, they suggested that only a catastrophic weather response could be expected to produce synchrony in a subnivean environment.

The types of weather perturbation that might significantly affect the direction of population growth emerged in the study of lemming cycles near Barrow, Alaska (Pitelka, 1973). Observations beginning in 1949 recorded peaks in 1953, 1956, 1960, and 1965. A snowstorm in early July of 1963 resulted in subnormal temperatures for the remainder of the summer. This was followed by excessive rainfall, producing general wetness and surface water extensive enough to limit both lemming and plant production; this may have postponed the expected 4-year peak to 1965. From 1965 to 1972, no typical peak (exemplified by high brown lemming numbers, extensive winter cutting of vegetation, and presence of predators) occurred. A prolonged storm in June 1965 produced a chaotic condition for breeding jaegers by making the abundant lemmings basically inaccessible, and the ultimate density of nesting jaegers was generally below earlier peak years: the resulting production of young was quite low. Snowy owl production was also low, so that the usual massive exploitation of the lemming population by avian predators did not occur; also, the pattern of predator dispersion, usually consistent as resident avian predators expand into territories abandoned by neighbors, was irregular. This probably allowed lemmings to survive in local patches and thus enter the postpeak winter in higher densities than would normally occur. By 1966, lemming numbers were low, but not as low as in typical postpeak years, and a slight recovery in numbers may have occurred. Numbers increased in 1967 and 1968. Evidence suggests that the population began an increase in the 1968–69 winter, but an ice mesh at the ground surface limited the accessible vegetation, resulting in low individual growth and low reproductive rates. The continued presence of unprecedented heavy weasel predation, building since 1967, combined with a severe winter in 1969–70 (a late autumn rain produced an ice glaze on the tundra surface, a situation called a flen-year by the Swedish Lapps, who associate it with the decimation of lemming hordes after a population peak; Marsden, 1964) reduced the number of lemmings to extreme scarcity in 1970. In 1971, brown lemmings, feeding on the by-then lush vegetation, were as high as in peak years, but appear to have mostly immigrated; and collared lemming (*Dicrostonyx*) were present in unprecedented quantities. Lemmings decreased during that summer, but the brown lemming population began to recover by late summer. By 1972, the numbers had again declined, owing probably to low winter snow cover and arctic fox predation.

This suggests that weather, although not providing any kind of timing mechanisms for cycles, can still have a major effect on the population. This effect may be the cause of the phase shifts that appear to occur between various continents and various areas on a given continent. This again emphasizes the local nature of these fluctuations. Unfortunately, lack of funding brought this study to a halt before the expected reestablishment of the cycle, and a pattern of recovery, could be observed.

If weather is to be postulated as a timing mechanism for the observed oscillations, it will have to explain two factors: synchrony and regularity. Poor weather might produce synchrony by strongly influencing peak populations while having little effect on increasing populations, and favorable weather might extend the duration of the peak in one population while accelerating the rate of increase in an expanding population (Krebs and Myers, 1974). Theoretical support for the possibility that weather could be a synchronizing force was demonstrated by Leslie (1959), who described a time-lag population model in which an external random factor (e.g., weather) could realign out-of-phase population densities in different areas; unfortunately, the supportive data to prove that this actually occurs are lacking (Krebs and Myers, 1974). It is also important to remember that a weather variable that operates over a short period (e.g., a single season) might be expected to align population peaks in adjacent regions, which on the contrary can show delays of several years. Perhaps a global factor, such as depression of mean annual temperature, which lasts for several years after major volcanic activity, could affect all the areas at a crucial point in the cycle (e.g., the low point). Watt's global temperature hypothesis (1969; see also Elton, 1924, 1930) is convincing from this standpoint, because it includes a long-term effect on global weather. However, this cannot in itself explain cycles observed and the phase differences among various regions, both distant and adjacent. Fitting an autoregressive process and calling the cycle endogenous does not explain why the coefficients of the series are such as to induce a pseudoperiodicity, or what prevents the observed autocorrelations from damping out. In environments subject to as many extreme fluctuations as the boreal forest and arctic tundra, it is difficult to understand how several species of fauna could assume population coefficients that result in regular periodic fluctuations of similar frequencies and different phase relationships over such vast areas.

Weather might be a viable synchronizing factor if one could postulate that similar weather conditions affect different phases of the population fluctuations in different ways (Frank, 1957). If the quality of the population varies with the density and if populations are sufficiently geographically isolated, adverse weather might severely affect peak populations while exerting little influence on expanding populations (Chitty, 1967, 1969). Whether or not this actually occurs does not appear to have been documented.

Another way in which weather might synchronize cycles in distant populations is mentioned by Keith (1974). The suggestion is that some weather sequences (e.g., two or more mild winters in succession) might allow high populations to persist at high levels, and a subsequent return to normal winters, combined with the heightened hare–vegetation interaction, would produce a widespread crash. This need occur only once every few decades, assuming a basic endogenous cycle. Since this remains apparently undocumented, it must be left as an interesting and convincing theory.

Resonance and Beats

Another possibility is that there could be a resonance effect between the properties of a biological community and some external force, which might be relatively inconspicuous (Hutchinson, 1942, 1949). The idea of resonance implies that a system has a natural, fundamental frequency of vibration; when an external vibration is applied to the system, it will only be maintained by force and in proportion to the energy applied, but when an external vibration approaches the fundamental frequency of vibration of the system, the system responds disproportionately and vibrates intensely at its fundamental frequency. Thus a wineglass can be made to respond to one precise note of a violin or a singer's voice. Any number of environmental variables, which might be considered as constituting a noise system, might resonate with the biological system at certain spectral frequencies (Hutchinson, 1954), and complex interactions among these variables might make several factors, seemingly unimportant when considered individually, more significant when operating together.

The basic idea of resonance phenomena is that the driving frequency of some external forcing function will produce large amplitude oscillations in a system if that system has a natural frequency of oscillation near that of the forcing function. This implies that (1) there is a natural frequency in the system slightly greater than the frequency of the forcing function, and (2) the resonance frequency, which is the frequency of the driving force, will be the observed frequency of the system. If we consider, for example, snowshoe hares and muskrats, we require a driving force with a period close to 9 to 10 years, plus a systemic population oscillation with a period slightly less than this. Since it is difficult to conceive of the latter for populations with life spans of individuals less than 5 years (and the same problem will exist for the lemming cycles), direct application of the resonance concept poses problems.

Another way of approaching the idea of resonance is advocated by Nisbet and Gurney (1976b). They interpret cyclic behavior as the response of a stable underdamped system (one that, in the absence of environmental noise, would approach equilibrium through a series of damped oscillations) to random noise. The amplitude of the oscillations would be a random variable, and the period would be determined by interactions within the system, with a steady loss of phase information due to the noise. Unfortunately, possible physiological and environmental factors that might produce such a system are not indicated.

However, the concept of resonance suggests another phenomenon, beats in harmonic oscillators. Beats are the result of two sinusoidal waves of similar amplitudes and differing frequencies which, as they shift in and out of phase, are occasionally exactly opposite, the nadir of one corresponding to the zenith of another. In sound, the waves would cancel each other, producing a beat (see fig. 72). The frequency of beats would be determined by the relative frequencies of the two sine waves. In this manner two cycles of quite different frequencies can produce a beat cycle (here, the observed population cycle) of yet another frequency. This is quite different from the idea of "tracking," where one cycle exactly follows another; or "cuing," where one cycle, which might be gradually shifting phase relations with other cycles, is brought back into phase by a major external shock, which might be cyclic, or merely occurs often enough to bring similar cycle periods back into phase (e.g., Keith, 1974).

Oster and Takahashi (1974), studying the effects of age structure and environmental periodicity on the behavior of interacting populations, found that by considering a population as a "distributed parameter system" with periodic forcing, periodic behavior might be observed as well as "distributed resonances," comparable to harmonic beats in linear oscillators. The state of an individual in a population at a particular time is specified by age since birth and a set of physiological parameters that reflects factors bearing on the individual's growth and reproduction rates. Each individual can be represented by a point in $(n + 2)$-dimensional state space (where n represents the number of physiological parameters considered), and it is assumed that knowledge of an individual's state is sufficient to predict its trajectory in the state space. To describe a population of these individuals, a set of density functions on the state

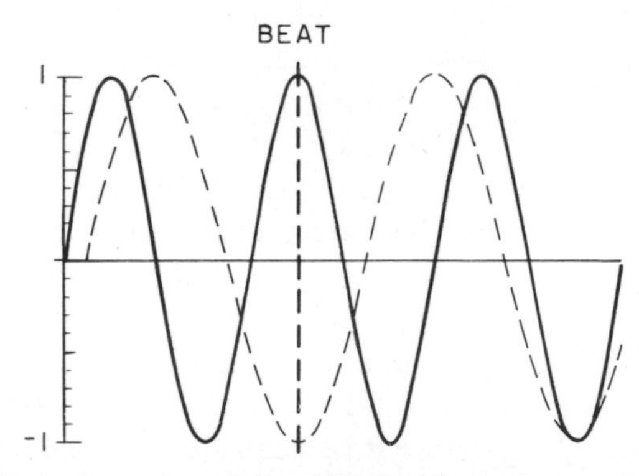

Figure 72. Visual representation of the concept of beats. As two sine waves of similar amplitudes but differing frequencies shift in and out of phase, they may occasionally be exactly opposite; that is, the zenith of one will correspond to the nadir of the other. When this occurs, a beat results. The frequency of these beats will be determined by the relative frequencies of the two sine waves.

space can be defined for each population. By applying a conservation law to each population, the equations of motion for population densities can be obtained. The birth- and death-rate functionals of a given population completely determine its numerical abundance. This model has a general form which, by appropriate assumptions on these functionals, can be shown to reduce to commonly studied population models such as the logistic model and Lotka–Volterra equations.

Using linearized versions of these model equations, Oster and Takahashi demonstrate that a periodic forcing of the birthrate alone will not produce a resonance in the population profile, but any periodic influence (in the form of a forcing function) in birth and death rates that operates in an age-specific manner can produce waves in the age profile which will be propagated through the profile until they reach the breeding period (assuming all reproduction to be at a single age) and will then provide positive feedback effects on births, producing a secondary wave. If the frequency of these waves is approximately proportional to a generation time, constructive or destructive reinforcement will produce a conspicuous periodicity in population numbers. Differential effects on maturation rates might also produce periodic effects. The generation time of the species and the frequency of the forcing function determine whether a resonance response will occur and what form it will take.

Similarly, Nisbet and Gurney (1976a), analyzing the population fluctuations of a species described by the logistic equation with a time delay added, found that the behavior of the system is unchanged by oscillations in the birthrate alone, whereas simultaneous oscillations in the birthrate and carrying capacity can have varying effects: if the population, in the absence of forcing oscillations, would normally be in a limit cycle, this cycle is usually unaffected by oscillations in carrying capacity at frequencies substantially different from the limit cycle frequency, but if the limit cycle frequency is close to the driving frequency (or a subharmonic of the frequency), the limit cycle may synchronize to the driver or its subharmonic. Large-amplitude oscillations in the birthrate can destroy the limit cycle but otherwise would have little effect.

This conceptualization can be extended to predator–prey systems. Again assuming that all reproduction for both predator and prey species occurs at a single age and is controlled by periodic environmental forcing functions, and expressing the intrinsic death rate for each species as a mortality factor dependent on the total population of the other species, the linearized total population responses can be shown to be characterized by (1) resonance terms related to age of reproduction and the proposed periodic forcing function affecting birthrate, and (2) harmonic oscillation terms related to inter- and intraspecific mortality rates (Oster and Takahashi, 1974). Since at least two sinusoids will be involved in the response of each population, harmonic beats will result at a frequency proportional to the difference between the birthrate resonance frequencies (1), which will be some integral multiple of the assumed periodic forcing function affecting birthrate and the interspecific interaction frequency. The significant aspect of these beats is that they can recur on a time scale longer than either the external forcing frequency or the life span of an individual in the population, since the beat period for two frequencies is inversely proportional to the difference between the frequencies. Thus population outbreaks and crashes can result as exogenous and endogenous waves reinforce and annihilate each other. Although this will be observed as an endogenous limit-cycle oscillation, the origin is the result of the distributed nature of the system and not the predator–prey interaction coefficients as in the Lotka–Volterra equations. As Oster and Takahashi explicitly state, the assumptions of their model limit its use for quantitative predictions, but the qualitative conclusions are apparently robust. Genetic or environmental variability producing growth, dispersion, or density-dependent migration might obscure the resonance spectrum. However, it appears that adding age-specific effects on abundance to generalized demographic equations can produce a periodicity that may exist on a time scale exceeding the life span of an individual in the population. Unfortunately, since introduction of actual numbers into this model might be naive, we once again have a presently unverifiable possibility.

What possibilities for beat phenomena between intrinsic population periodicities and extrinsic periodicities could exist? If the period of the external factor is C_E and the period of the internal factor is C_I, the period C_B of observed beats will be

$$C_B = \frac{1}{1/C_E - 1/C_I}$$

We note that for a given value of C_B, if C_E becomes quite small,

$$C_I = \frac{C_B C_E}{C_B - C_E} \to C_E \qquad \text{for } C_E \to 0$$

At this point the system becomes too sensitive to make any decisions based on the crude biological estimates available to us. Similarly, as C_E becomes quite large, with C_B constant,

$$C_I \to C_B \qquad \text{for } C_E \to \infty$$

so beats associated with rather long term periodicities would be indistinguishable from tracking or cuing. Therefore, we consider external periodicities of an intermediate range for $C_B = 4$ years and $C_B = 10$ years.

What forms of external cycles could produce beats? First, what types of external cycle periods can be considered? We might assume that a beat phenomenon would act within the life span of an individual, since otherwise an internal cycle would have to be interpreted as a density-dependent social interaction, which undoubtedly exists but probably not with the requisite regularity for beat phenomena. For the average lemming life in Alaska of 1 year, the maximum for the external cycle would be on the order of 1 year. For the average snowshoe hare lifespan of about 5 years, the maximum external period would be less than 4 years.

The difficulty is finding an external factor that evidences regular cycles of an appropriate duration. The 11.2-year sunspot cycle, discredited as a driving force for population cycles since the period is too long to allow sunspots to act directly (Elton, 1931c), is also too long to induce beats at the required frequency. Rowan's (1950) ultraviolet radiation hypothesis, with ultraviolet intensity mediated by the earth's ozone layer, is in the same category.

Siivonen and Koskimies (1955) suggested that the 29.5-day lunar cycle might be involved. They observed that the 10.9-day difference between the lunar and solar years results in the shifting of the moon phases with respect to a fixed date on the solar calendar. If a fixed "critical period," during which reproductive success might be adversely (or positively) affected, should occur, say, in the spring near the beginning of breeding, this could generate periodicity, depending on whether or not the specific lunar event occurred at that "critical period." Assuming a new moon to be the event, a 9.6-year average (8- to 11-year) cycle can result. If the critical period is lengthened to about 6 days, the 4-year cycle results. The problem with this theory is that given the sizable environmental and individual variability in ecological systems, it is difficult to believe in the existence of critical periods of so brief a duration. And in terms of harmonic beats, this is in the category of $C_I \rightarrow C_E$ for C_E small, so that defined period lengths could be somewhat arbitrary. This theory seems to provide a good example of "how some data may be manipulated to produce nonsense correlations" (Keith, 1963).

However, other planetary cycles might usefully be investigated. For example, for a 10-year cycle, the period of the conjunction of Venus and Mars (0.914 year) forms a beat with an internal cycle of 1 year (table 6), which might represent annual breeding. However, a similar correspondence is not apparent for 4-year cycles. Snowshoe hares and lemmings

TABLE 6 Beat Frequencies

External Factor	C_E	C_B	C_I
Jupiter – Venus conjunction	0.65	4 yr/10 yr	0.78/0.70
Mars – Venus conjunction	0.914	*Lemmus* life span	1.18/1.01
Earth – Venus conjunction	1.599		2.66/1.90
Mars – Earth conjunction	2.14		4.60/2.72
Lunar nodal apside	2.9985	*Lepus* life span	11.98/4.28

Note: C_E, period of external factor; C_I, period of internal factor; C_B, beat frequency.

could form harmonic beats with different external cycles of differing origins, but this seems to enter the realm of valiantly seeking numbers that fit, and since there are as yet no observations to support this hypothesis, it will not be pursued here in greater depth.

Another possibility that has been suggested (Collin, 1954; May, 1976b) is that populations might track external cycles. For example, one might observe that six periods of the conjunction of the Earth and Venus with the Sun give a period of 9.6 years, the average period length recorded for the 10-year cycle in lynx. Other conjunctions of this period length also occur (Collin, 1954). For the 4-year cycle, 2.5 periods of the conjunction of Venus with the Earth, and 50 lunar months, both give periods of 4 years. These observations are interesting and seem worthy of further consideration, but one rather important factor is difficult to directly explain by a cosmological driving force: if, for example, lynx pelt data are indicative of relative population numbers, the peak populations in various parts of Canada, assuming a cosmological factor by itself, would be expected to be synchronous, which does not appear to be the case (Elton and Nicholson, 1942b). Subjective examination also suggests that phase shifts occur between similar fox species in Labrador and Norway. However, the concept of a regular external timing mechanism, such as planetary movements, may prove to be useful.

STRESS

Green and his coworkers (1938–40), working with snowshoe hares in Minnesota, suggested that a syndrome referred to as "shock disease," characterized by low blood sugar, degeneration of the liver, and failure to store glycogen, might be the cause of the periodic decimation of snowshoe hare populations. Although their interpretations of the "shock syndrome" were not actually supported by their observations, and no observations of this syndrome in natural populations have been recorded (Chitty, 1959), their description of shock disease stimulated Christian (1950) to extrapolate from the work of Selye (1946) on the response of the pituitary and adrenal glands to stress due to lack of food and cover, intraspecific conflict, and increased predation pressure, and to formulate what is now called the *stress hypothesis*. Specifically, Christian proposed that excessive stimulation of the adrenopituitary system resulting from increased stress at high population densities, especially in winter, followed by late winter demands on the reproductive system stimulated by increased day length and other factors, can produce exhaustion, low resistance, and susceptibility to mortality factors. Concurrent intensification of social aggressive behavior might result (Christian, 1975). High density should be thought of not as absolute numbers, but as pressure on the upper limits of the environmental carrying capacity so that varying amplitude of population peaks can be seen as temporal variations in carrying capacity. Thus Christian (1950) suggested that beaver do not cycle because heavy trapping pressure keeps their population below the carrying capacity of the environment. Similarly, muskrats cycle except in areas where water levels are controlled and trapping pressure is intense, although even these populations can crash if adverse environmental circumstances lower the carrying capacity (Christian, 1950).

Krebs's (1964a) observation on *Lemmus trimicronatus* and *Dicrostonyx groenlandicus* at Baker Lake, Northwest Territories, were reexamined by Christian (1975), who showed that adrenal weight at the population peak was higher than before or after the peak for both species. Andrews's (1968, 1970) studies on *L. trimicronatus* at Barrow suggest a higher adrenal secretory rate in natural "high-density" populations with higher secretory rates in winter than in summer; however, varying time spans between sample collection and final analysis leaves these results open to question (see Krebs and Myers, 1974, for a general discussion of this problem). The fact is that the extent to which increased adrenopituitary function plays a role in causing mortality in natural populations is essentially unknown (Krebs and Myers, 1974; Christian, 1975). It seems likely that death rates from predation and other causes are so great that increased mortality from endocrine activity will be negligible (Christian, 1975). It is more likely that reproductive changes induced by increased adrenopituitary function will be a limiting factor. Evidence suggests an inverse relationship between the functioning of the pituitary–adrenal axis and the pituitary–gonadal axes in rodents (Brain, 1971): this could account for a shortened reproductive season or decreased reproduction rate in years of high density. It has also been suggested that maturation rates might be density-dependent (Christian, 1975).

One significant problem with stress theories is the elusive nature of stress. G. E. Hutchinson (personal communication) noted that "stress" usually refers to "something that I am talking about but do not understand." Another major difficulty with stress theories in the light of population cycles is that peak amplitudes vary enormously. As stated previously, "density" would be affected by carrying capacity, so that "stress" may be more related to limitations of forage than to large population numbers or aggressive encounters. This is supported by the fact that high "density" (related to *area*, not carrying capacity) does not necessarily produce a decline (Chitty, 1960). Another argument against stress as a primary driving force is that observations on a cycling vole population in Massachusetts showed no significant increase in adrenal weight (as a measure of stress) concomitant with an increase in population size (although it is not clear that the peak of this population over the 3 years of observation actually reflects an extremum for the carrying capacity) while a noncycling vole population (*M. breweri*) on a nearby island did evidence a high level of stress (To and Tamarin, 1977), probably as the result of the lack of an outlet for dispersal (Tamarin, 1978). It seems likely that stress and dispersal are generally interrelated. This idea is supported by the observation that the postcapture mortality (used as a measure of stress) was 50 percent in brown lemmings at Barrow, Alaska, during a population high in 1971, compared to a mortality of 0 to 1 percent in other years (Andrews et al., 1975), and it is likely that this population was mostly migrants (Pitelka, 1973).

In snowshoe hare research, an unpublished study by Feist and coworkers at the Institute of Arctic Biology in Fairbanks, Alaska (discussed in Wolff, 1977a), showed no correlation between physiological responses of the neuroendocrine system and snowshoe hare density, and Windberg and Keith (1976a) found no evidence of hare social stress as a direct mortality factor in artificial high-density populations.

Although stress may be connected with variations in the size of populations, it is not immediately obvious how this factor could by itself produce the type of regularity suggested here. Thus endocrine–behavioral negative feedback mechanisms may operate to increase mortality and decrease reproduction, with increased density in some small mammals, and may by this means enhance the amplitude of population fluctuations, but they would not appear to be factors producing regularity in these oscillations.

Genetic–Behavioral Polymorphism

Chitty (1957, 1970) proposed that deterioration in the quality of a population as a result of selection on genetically determined behavioral types might temper population increase. He suggested specifically (Chitty, 1967) a change in the genetic composition of the population during the cycle, with natural selection favoring large aggressive animals with low reproductive rates at high densities, and smaller, less aggressive animals with high reproductive rates at low densities. This fits the observations that declining populations of *Microtus agrestis* are composed of some individuals with high growth rates and others with low growth rates (Newson and Chitty, 1962) and that the relationship between skull and body measurements in brown and varying lemmings is associated with changes in density (Krebs, 1964b). Spacing behavior, measured as aggression, would be the force driving the demographic machinery through the cycle. Studies using electrophoretic variants as genetic markers demonstrated dramatic genetic changes related to density changes (e.g., Myers and Krebs, 1971; review in Krebs and Myers, 1974; Krebs, 1978). However, these electrophoretic genetic systems may not be valid indicators of demographically relevant selection (Kohn and Tamarin, 1978; in Tamarin, 1978), and there is as yet no conclusive genetic evidence to support Chitty's genetic–behavioral hypothesis in microtine populations (Krebs and Myers, 1974).

Is there any evidence for changing behavioral interactions during a population cycle? Working with male *Microtus pennsylvanicus* in Manitoba, Turner and Iverson (1973) found that aggressive behavior increased during the breeding season and was enhanced by decreasing size of the home range, but they could not find a consistent relationship between level of aggression and population density. Smaller home ranges and larger animals, two characteristics of peak populations (Krebs and Myers, 1974), appear to be related to male aggressiveness (Turner and Iverson, 1973), but since fighting is rarely lethal, it is not clear how these behavioral interactions could account directly for mortality changes in declining populations. There is as yet no indication that these mechanisms are part of the snowshoe hare cycle (Wolff, 1977a). It seems more likely that this hypothesis, and the stress hypothesis, will be subsumed by the larger concept of spacing behavior and dispersal.

Spacing Behavior: The Role of Dispersal

The famous mass migrations of Norwegian lemmings clearly indicate that dispersal plays a role in the cycle scene. The spectacular migrations undertaken by arctic foxes, usually connected with lemmings crashes, make the role of dispersal even more emphatic (Braestrup, 1941; Banfield, 1974). However, mass movement may be more a local response, either to a bad situation or to the possibility of establishing populations elsewhere, than a cause of regular periodicity. The local nature of lemming fluctuations in Norway was emphasized by Collett (see Wildhagen, 1952) and Elton (1942), who independently observed that mass migrations occur *somewhere* in Norway on the average every 3 to 4 years. Collett (1895) also noted that, although migrations appear more frequently around the Trondheim Fjord (Nord Trøndelag) than in other areas of Norway, minor exoduses may often include only a few valleys adjacent to the optimal plateau. Thus a lemming migration peak does not mean a resultant migration. Collett (1911–12) added that peak years could sometimes include vast areas of Norway, but he did not mention whether widely separated local populations might be generally synchronous. Pitelka's (1973) observations in Barrow, Alaska, also emphasize the local character of lemming flux. Schultz (1969), also working at Barrow, noted that even during years of lemming lows one could find "pockets" supporting a denser population: as an example, the fluctuations in lemming numbers on the outskirts of the Barrow Eskimo village are not as pronounced as in the open tundra; and 100 miles east of Point Barrow the lemming population peaked in 1957, a year after the Barrow peak, but was in phase again in 1960. Cross (1940), in his study of periodic fluctuations in the red fox (*Vulpes fulva*) in Ontario, observed that periods of maximum populations were regional and not province-wide and that fur-trading posts that did reach maxima in the same year usually occupied a contiguous territory. Elton and Nicholson (1942b) discovered differences in timing of peaks for lynx populations in widely separated areas, and hare fluctuations are similarly regional phenomena (Seton, 1909); this means that a period of continent-wide abundance might cover several years.

If these fluctuations are essentially local, what is the role of dispersal? It may be a means of escaping adverse circumstances created by high-density and/or food shortage—while also providing a means for colonization. The essential feature of lemming mass movements is probably not mortality but the survival of some of the lemmings and the expansion of territory (Kalela et al., 1961). This is implied in Collett's (1911–12) observation that the "migratory instinct" may basically represent the desire to follow one's neighbors as they move, and that if individuals become isolated from others, the desire to migrate may diminish and the lemmings may colonize. Alternatively, it seems quite plausible that migrations may begin as a seasonal change of habitat from the preferred winter habitats in the moss-rich alpine zone under a protective snow cover, to the favorite summer peatlands in the birch zone and lower parts of the alpine zone (Koponen et al., 1961). In many years, spring

Norwegian lemming migrations do not go over the closest fjelds, and lemmings may stay and reproduce in the higher parts of the pine forest. In "great" years lemmings descend the mountains through the forest belt either individually or in groups, their direction basically determined by the landscape, and soon they arrive at a forest or valley. They may stop for the winter, but in peak years they may spread across the valley or even cross large bodies of water to try to reach distant land. Although the lemmings may return to former sites if they only proceed as far as shallow valleys, once they reach the bottom of large or open valleys they never return to the high fjelds (Collett, 1911–12). Thus what begins as a common seasonal relocation in search of fresh forage areas and lower population densities may become a confused dislocation, or a continuing search for acceptable territory. Collett (1911–12) definitely intimates that the lemmings are looking for something in addition to leaving something.

Pitelka (1973) suggests that a similar phenomenon occurs in Alaska. In 1971 he recorded a mass winter ingress of brown lemmings into a thriving vegetation in which the local brown lemming population had remained low; these migrants appear to have been suffering from some type of severe stress, since the postcapture mortality was much larger than in other years (Andrews et al., 1975). The fact that the following autumn most of the migrants were gone suggests that they did not find what they were seeking. This is not surprising, because, although the collared lemming population had reached an unprecedented high, the local brown lemming population had not recovered from its earlier crash. When food becomes limiting in snowshoe hare populations, dispersal seems to become a key factor: Windberg and Keith (1976a), creating an artificially high density of 29 hares per hectare in a confined island population of hares, noted the establishment and continuous use of runways along the island periphery, and they interpreted this as a motivation to leave the island before exhaustive overbrowsing of island vegetation occurred. Highest rates of movements appear to be associated with peak population densities (Windberg and Keith, 1976b). The first habitats to be vacated during the decline phase are the open marginal habitats, then suboptimal habitats, and finally almost all habitats. Hares select habitats on the basis of food supply and protective cover, utilizing the most dense habitats first during population expansion, and abandoning the least dense habitats first during population declines (Wolff, 1977b). Snowshoe hares apparently can also be subject to mass migrations, as observed by Cox (1936) and Henshaw (1966). The fact that observed migrations occurred during peak hare densities suggests that the migrations were the result of frustrated dispersal (Wolff, 1977a).

The emphasis on the local nature of population cycles, combined with the significance of dispersal, is reminiscent of a model developed by Anderson (1970) based on observations of the house mouse, *Mus musculus*. Following the direction of earlier Soviet investigators, Anderson emphasized a dichotomy in local habitats, classifying them as "survival stations" and "colonizing stations." A survival station is a locale continuously occupied by a particular species; density and population composition are relatively stable and inbreeding is the rule. A colonization habitat is characterized by intermittent or seasonal occupancy and extreme fluctuations in density. In periodic species colonization habitats are unoccupied during low population years, but may be sites of maximum density in peak years. Population eruptions may occur because of the varying carrying capacity of the colonization habitats. Coincidence of environmental conditions may result in major regional increases. Kalela's (personal communication; Anderson, 1970) observation that *Lemmus* highs in Finnish Lapland occurred following occupation of secondary breeding areas supports this view.

This model seems consistent with what is known about population cycles. For example, Krebs et al. (1969) suggested that dispersal might be necessary for population control in voles, since fenced populations in Indiana were unable to control their population density below starvation level. Also, Krebs et al. (1976), after clearing several areas near Vancouver, British Columbia, to study the colonization pattern of *M. townsendii*, found that increasing populations tended to lose individuals due to dispersal rather than to death; high-weight voles, which tend toward aggressive behavior (Turner and Iverson, 1973), were characteristic of late increase and peak populations, but few of these dispersed. A more complete picture is drawn by Lidicker (1973) for a population of *M. californicus*: animals were only found in refuges during population lows; the density in the refuges did not increase significantly as the population grew and dispersed into surrounding habitats, but when all suitable habitats reached their carrying capacity, the population demonstrated frustrated dispersal (i.e., a tendency to disperse with nowhere to go).

In his studies on snowshoe hares, Wolff (1977a) described a similar situation:

> During the low phase of the hare cycle, hares were found only in the refuges. During high population densities, hares were found in all suitable habitats.... The demographic consequences of dispersal are dependent on the relationship between population densities and carrying capacities when the dispersal is occurring. In this study it was found that hares dispersed from the refuges into suboptimal habitats during population growth before the population reached its carrying capacity.... Suboptimal habitats frequently have more

food seasonally, but there is less cover and hares are more exposed to predation. . . . The predation rate is low even in these more open areas due to the lack of predators during the low phase of the cycle. With a high reproductive rate (as many as 15 yg/female/yr) the population in both optimal and suboptimal habitats increases. Presaturation dispersal (defined below) occurs until all habitats are filled. The population density within the optimal habitats does not increase strongly but rather surplus animals disperse into suboptimal and marginal habitats.

Lidicker (1975), in a general description of small mammal demography, used the term "dispersal sink" as an alternative to what Anderson calls a colonization habitat. Dispersal from a survival habitat (or refuge, or optimal habitat) in which the population is below the environmental carrying capacity, was described by Lidicker as "presaturation dispersal." This type of presaturation dispersal into a dispersal sink is probably what Collett (1911–12) was describing when he noted that in many years, lemming migrations in the Norwegian mountains did not even proceed over the closest fjells, and many lemmings would become established in the upper parts of the pine forests and reproduce there; but in favorable years, which may sometimes be associated with early reproduction in spring, with the first litter then being reproductively active in August, lemmings would go down the mountains through the forest belt, *either individually or in groups*, stopping either in the forest or the valley or at the coast or even swimming across a body of water to a body of land that is in sight.

Tamarin (1978) has suggested that the presence of a dispersal sink is requisite for cycles to occur, and he has developed a continuum model in which the degree of cycling is related to the degree of dispersal; he connected this with the concept of the $r - K$ continuum (MacArthur and Wilson, 1967; Pianka, 1970). This idea emerged from the observation that apparent cycles in *M. pennsylvanicus* on Cape Cod were not found in *M. breweri* on a nearby island after 5 years of study (Tamarin, 1977): the island population tended to maintain a high density with no apparent dispersal sink (no place for migrants to go, and too little predation to produce a steady loss) and thus would be expected to be under K selection (i.e., the carrying capacity is the limiting factor), whereas on the mainland, where the population experiences declines and possibilities for expansion, r selection (where the inherent rate of increase is the determining factor) should occur at least part of the time. During a cycle, shifts along the r–K continuum could occur as density increased. In vole populations, high dispersal occurs during the increase phase; eventually, the dispersal sink could become saturated, and this saturation could be especially pronounced if several local populations in a given area were to become synchronized. If predators became saturated, the dispersal sink would be minimal and a shift toward *K* selection could result, as evidenced by changes in body weight (e.g., Chitty, 1952) or other parameters.

This description corresponds to Pitelka's (1973) description of lemming cycles at Barrow, Alaska, Krebs' and Myers' (1974) description for small mammal cycles in general, and Keith's (1974) description of cycles in snowshoe hares. Additionally, many of the varying parameters attributed to the cycle "syndrome" (e.g., variations in body weight, skull–body regressions, stress, juvenile growth rate, early testis regression or early cessation of reproduction by females, etc.) might be viewed in this context as responses to high density, with that parameter responding to selection most conspicuously which responds most readily (Schaffer and Tamarin, 1973).

It would be quite interesting to test Tamarin's hypothesis by looking at other island populations. For example, the Tunnermeriut Eskimos on Bylot Island in the North American Arctic, an island where *Dicrostonyx* is much more common than *Lemmus*, say that they have never seen a lemming migration, and no migrations have been noted in their folk tales (Miller, 1955a). On the other hand, Collett (1895) noted that on several large islands and "even quite insignificant islets" off the coasts of Trømso in north-central Norway, Norway lemmings could suddenly be observed in swarms migrating in the autumn down to the shore, and this could occur when no population high was observed on the adjacent mainland. A study of the biogeographic structure and predation in these areas might validate, or clarify, the idea of a dispersal sink as applied to oscillatory species.

Tamarin (1978) also suggested that "if declines occur on the mainland because the gene pool is changed by dispersal, one would not expect a cycle on Muskeget [Island] where the gene pool is not being changed by dispersal. If the genes we are talking about are related to behavior, then we are supporting Chitty's (1967) behavior–genetic polymorphism hypothesis." But there is an alternative type of "genetic change," proposed by W. D. Hamilton (1971) and independently by E. L. Charnov (personal communication; see Charnov and Finerty, 1978) which could be considered in this same context. The general pattern for cyclic microtines (Anderson, 1970; Lidicker, 1973) and snowshoe hares (Wolff, 1977a) is to persist mainly on restricted islands of favorable habitat (in the sense of Wright, 1943) or survival stations (Anderson, 1970) during years of population lows, and then to disperse into less favorable habitats (usually determined by quality of forage and/or exposure to predation) as numbers begin to increase. At a population low, when there is little dispersal, individuals interact with those in their neighborhood, who are probably close relatives (Hamilton, 1964, 1971, 1972). That adjacent local populations may be genetically dissimilar has been demonstrated (Gaines and Krebs, 1971).

The coefficient of relationship, which is an index of the expected degree of altruism and selfish behavior among individuals, is on the average high between individuals when the population is low. During the breeding season aggressive behavior may increase, as observed in male *Microtus* in Manitoba, and this aggression may be enhanced by decreasing size of home range as the population increases (Turner and Iverson, 1973). As local populations increase, this aggressive behavior may initiate dispersal: in fact, Clough (1968) suggests that this may describe the initiation of mass movements in Norwegian lemmings, and the observation that isolated lemmings will cease migratory behavior (Collett, 1911–12) is consistent with this idea. At this point a dispersal sink, which may include colonization habitats, predators, and so on, is available; if a dispersal sink is lacking or inadequate, the population may stabilize relative to the carrying capacity of the environment and not cycle (Tamarin, 1978). With increases in population and dispersal, migrant individuals entering a neighborhood (where they are probably unrelated to the residents) will cause a drop in the average coefficient of relationship—during high dispersal this drop will be rapid. Concomitantly, an increase in aggression among less-related individuals could be expected. There is some evidence that microtines in peak populations are more aggressive (Krebs and Myers, 1974), and although Turner and Iverson (1973) did not find a consistent relation between aggressive levels and population density in their natural populations, they did find a correlation between aggression and increased body size and decreased home range, both features of peak populations. The results of these behavioral interactions may produce general individual weakness through stress (see Andrews et al., 1975) rather than direct mortality.

The genetic change described here is the relatedness of individuals in a given geographic location and is brought about by the dispersal itself—it does not require that the dispersers be of a different genotype than those which remain at home (which may be the case; see Myers and Krebs, 1971). As long as dispersal is high and the average coefficient of relationship is low, aggression should remain elevated, but dispersal would probably stop when the carrying capacity of the environment, which might vary from cycle to cycle, is approached. If "high population density is not sufficient to produce a decline, and low density is not sufficient to stop a decline" (Krebs and Myers, 1974), it may be because we err by measuring density as numbers per unit area rather than numbers with respect to current environmental carrying capacity (including area, food supply, predation pressure, etc.) and in particular the carrying capacity of the dispersal sink. Consequent environmental depletion of the dispersal sinks would suggest that losses by emigration might be proportionally highest in the increase phase and lowest in the decline phase, as observed by Myers and Krebs (1971).

When dispersal has ceased, a period of time will be required to restore relatedness in local populations. Intensified behavioral interactions during this period may cause a population decline directly, for example by affecting growth and reproduction, or indirectly, by making individuals more susceptible to predation, extremes of weather, or disease (Krebs and Myers, 1974). Such responses might be sufficient to provide the steady, small mortality factor necessary to account for most losses in declining populations: Krebs and Myers (1974) estimated that a 10 to 15 percent drop in the probability of survival per month would be adequate for most microtine declines. Declines would be attributable to intraspecific aggression; thus it would not be inconsistent with this theory to observe *M. ochrogaster* suffering high mortality and numerical decline at a time when *M. pennsylvanicus* was surviving successfully in the same Indiana field (Krebs et al., 1969); or to note that an increased population of *Lemmus lemmus* in Finland occurred simultaneously with a decreased population of *M. agrestis* (Tast and Kalela, 1971). Declining populations should result in quick withdrawal from marginal habitats (consider Keith, 1974, and Wolff, 1977a, for snowshoe hares; Tamarin, 1978, for voles), reestablishing local islands with an increasing average coefficient of relationship. Thus the stage would be set for another population cycle.

The Tamarin model and the Hamilton–Charnov model are essentially the same, the basic difference being the nature of the genetic change. There is evidence, however, that this difference may not exist. Garten (1976), in a study of oldfield mice (*Peromyscus polionotus*) at sites in Florida, Georgia, and South Carolina, examined heterozygosity at 31 protein and enzyme loci using electrophoretic techniques. He discovered that in mainland populations a high correlation exists between the percent of heterozygosity in the population and the ability to compete, as measured in an arena by the number of encounters resulting in fighting, decrease in fight latency, mean social dominance, and increased body weight. This supports the relationship between population dynamics and genic heterozygosity suggested earlier by Smith, Garten, and Ramsey (Smith et al., 1975). While these observations are generalized for whole populations and do not refer to individuals, they may help to clarify the idea of a genetic-behavioral mechanism by suggesting that behavioral vigor is associated with the proportion of heterozygous loci present—thus making the mechanism dependent on heterotic effects across the entire genome rather than linking it to a small number of alleles (Smith et al., 1975). Relatedness among individuals is clearly implicated here because inbreeding during population increases following the cessation of large-scale dispersal would tend to decrease heterozygosity, thus decreasing aggressiveness and reproductive rate (Smith et al., 1975); assortative mating and genetic drift might also decrease heterozygosity (Garten, 1976). It may be that genic heterozygosity will link the Tamarin and Hamilton–Charnov theories as one. The critical test will be to demonstrate in the field that behavior, population numbers, and genic heterozygosity at the *individual* level changes concurrently, and whether these electrophoretic genetic systems are actually valid indicators of demographically relevant selection which, as noted earlier, has been questioned (Kohn and Tamarin, 1978; in Tamarin, 1978).

This model is appealing because it finds places for essentially all of the previous observations that have been made in cycling species (viewing most as results and not causes), and it allows for a situation in which a normally cycling species might not cycle, namely absence of a dispersal sink. Tamarin's (1978) study and several others (refer to Tamarin, 1978) suggest that this may work for cycling microtines—but does it work for cycling lagomorphs?

The apparently anomalous factor in the snowshoe hare cycle is the so-called north-south gradient in hare cycles: snowshoe hares cycle in the northern part of their range, but not in the southern part (which includes primarily the western mountainous region of the United States). This is more accurately expressed as decreasing amplitude of fluctuations as one moves south. Howell (1923), Leopold (1933), and others (see Wolff, 1977a) thought that this fact might be related to the generally scattered and discontinuous pattern of the spruce–fir hare habitats in the southern regions. However, during low population densities, hares in Alaska are basically confined to about 10 percent of their suitable habitat, which may be comparable to hare distribution in the western United States. In both Alaska and the western United States, increased hare density leads to dispersal into less favorable habitat, but in Alaska hares become established in less favorable habitat (Wolff, 1977a), whereas in Colorado, for example, they do not (Dolbeer and Clark, 1975). This may be due to differences in predation, since the major hare predators in the north (lynx, goshawks, and great horned owls) are obligate migratory predators displaying a density-dependent cycle with hares so that during hare population growth, predation in exposed, less favorable habitats may be no more than in optimal habitats; whereas in the western United States the hare predators (red fox, coyote, bobcat, several species of hawks and owls) are facultative, resident, and noncyclic, with the result that they exert continuous pressure on hares moving into less favorable habitats.

What this implies is that hares in the southern part of their range do not cycle because of a *highly effective* dispersal sink, which would appear to contradict the Tamarin model.

One possibility is that these populations do cycle, but at a much-reduced amplitude. This may be suggested in the work of Windberg and Keith (1978) on snowshoe hare populations in Alberta woodlots, which are patches from a former forest that are now isolated in an extensive agricultural area. In this study the researchers found that population trends in woodlots were similar to those in the nearby forests, except that the amplitudes of the flux were much reduced (relative high/low = 6 : 1, compared to 35 : 1). They attributed the restrained density peak in the woodlots to low rates of first-year juvenile survival and possibly higher predation rates, but it may be that there is heavy predation in the marginal habitats that has not been observed (Tamarin, 1978), and that the situation is directly comparable to that in Colorado, so that unnoticeably low peaks may be occurring in Colorado and other southern regions. Alternatively, the southern populations may not cycle at all, and it may be that the concept of dispersal sink in the model will have to be reworked. This possibility will be explored later.

Epicentric Migration

The effects of prey dispersal are indirectly reflected in the mass movements of predators. For example, a large part of predation on lemmings is by migrant avian invasion, and this is enhanced by the influx of mobile mammalian predators (Pitelka, 1973). Mass movements in lynx, horned owls, and goshawks have long been assumed to follow regional hare declines (Hewitt, 1921; Keith, 1963), and several studies have supported this assumption (Keith et al., 1977).

A study of the color-phase genetics of colored foxes (Butler, 1945, 1947, 1951) offers evidence that migration is a major phenomenon for this predator species, too. We recall that there are three phases of the red fox: red, cross, and silver. Data presented by Elton (1942) revealed a steady decrease in the percentage of the silver phase in Labrador from 1834 to 1933. A similar decrease was recorded in British Columbia and the Mackenzie River basin. This progressive decrease continued from 1916 to

1944 while the general population increased sharply (Butler, 1945). It is known that the three color phases result from a single pair of alleles, homozygotes being red (dominant) and silver (recessive) and heterozygotes being cross. At least two different mutations have occurred, giving rise to a "Canadian gene" in eastern Canada and an "Alaskan gene" in western Canada (Butler, 1945). Factors shown to affect the relative proportions of the three color phases are (1) locality, (2) state of the population cycle, (3) population trends, and (4) migration pressure. Selection could be invoked to explain (1) and (3), but cannot adequately explain fluctuations of proportions during the cycles. Migration, however, with or without selection, can explain all four criteria, since migration produces a mixture of resident and migrant populations with different gene frequencies, and migrations would be expected to be rhythmic in connection with cyclic population flux. In one particular migration year in the Mackenzie River area, pelt returns revealed aberrant color-phase ratios, but these approached an equilibrium the following year; the equilibrium could be either at the old ratio or a new one, depending on the success of the migrants in establishing themselves as members of the local breeding population. Rapid population increase in an area of low silver fox frequency followed by migration from this area into areas of higher silver fox frequency would account for the general diminution in the percentage of silver and cross foxes in the populations. Some areas showed no correlation between the percentage of red foxes and the 10-year cycle, but did show a definite 4-year cycle in reds, which could also be explained by migration, in this case from the tundra regions to the north of the Mackenzie River locale, where foxes evidence 4-year periodicity (Butler, 1947).

A similar study in Quebec (Butler, 1951) supports the latter view. Records in the Hudson Strait region (which is predominantly tundra supporting a prey population of collared lemmings (*Dicrostonyx*) and mice, with some arctic hare but no snowshoe rabbits) from 1916 to 1948 suggest that a long-term cycle in colored fox coexisted with a 4-year cycle in arctic fox until about 1930, after which the 4-year cycle began to dominate in the colored fox population. This can easily be explained by assuming that the long-term cycle is a record of overflow from adjacent population peaks in forest regions, and that the amplitude and duration of these peaks depend on (1) availability of quality food in the tundra region, (2) amplitude of peaks in neighboring forest regions, and (3) direction of migrations out of heavily populated regions.

Trends in predator movements have suggested a theory in which an epicenter region (one might presume this to be in some respect(s) an optimum habitat) generates a rapidly increasing population, which feeds in a wavelike manner into neighboring areas. Elton and Nicholson (1942b), in their classic study of the lynx population cycle, showed such a broad trend in lynx population peaks outward from the Athabasca River basin in northern Alberta, Canada, and terminating (i.e., appearing last) in the Great Lakes and Gulf areas in eastern Canada, a trend that was suggested to Elton by Rowan. A recent time-series analysis by Bulmer (1974), examining phase differences between various regions, corroborated the finding of Elton and Nicholson for the lynx in the nineteenth century, showing that the fur returns peak earliest and most strongly in the Athabasca basin region (fig. 73), becoming weaker as they move away from that area (Bulmer also suggested that this appeared to be true in the twentieth century for lynx, red fox, and fisher). A similar picture is suggested for snowshoe rabbits and mink (Butler, 1953; fig. 74). An examination of color phases in Canadian foxes (*V. fulva*) revealed extensive movements across the continent, originating in the Athabasca region (Butler, 1947, 1951, 1953). Mapping of data on snowshoe hares from questionnaires suggests that the rabbits peak first at an epicenter region in the northern part of the prairie provinces and spread gradually

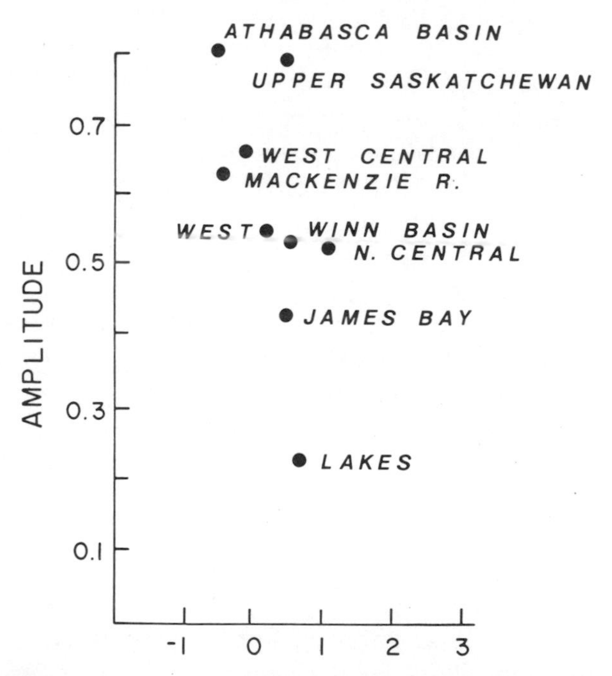

Figure 73. Relative phases of lynx from all of Canada, 1848–1909. Fur returns peak earliest in the Athabasca basin region and peaks are strongest there, becoming weaker as they move away from that area. (Based on Bulmer, 1974.)

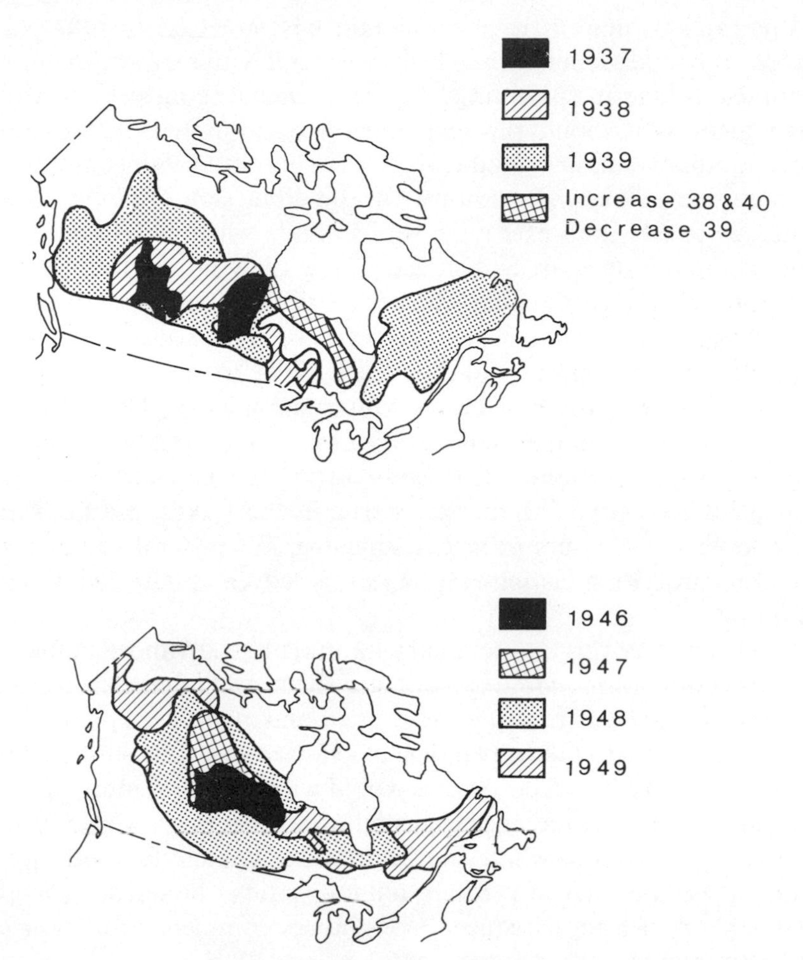

Figure 74. Phase relationships of snowshoe rabbit populations in Canada. Top: isophasal lines delineating areas of Canada where observers reported increases in rabbit population, 1937–40. Bottom: isophasal lines delineating areas of Canada where observers reported increase in rabbit population, 1946–49. (After Butler, 1953.)

outward over a period of 3 to 4 years (Chitty, 1950; Keith, 1963). Colored fox data from Canadian questionnaires are said to follow the same pattern, except in the area around James Bay and Hudson Bay, where unconnected increases occur. These unconnected increases can possibly be explained by the fact that, as is obvious from the Labrador data, coastal points are affected by the 4-year tundra cycle. Additionally, we have seen that the data for Norwegian foxes suggest this idea of an ideal habitat feeding into neighboring environs. In both areas the direction and extent of the migrations would clearly depend on population pressure and the current suitability of the habitat encountered.

How can this epicentric phenomenon be connected to other observations? It would seem reasonable to connect it with renewed availability of suitable forage in suboptimal habitats, encouraging prey to expand from refuges. Why should this phenomenon occur in the Athabasca basin before another locale in Canada? It may be connected with climatological considerations. As noted previously in the discussion of boreal climates (cf. figs. 59 to 64), the western areas of Canada warm before the heavily snow-covered eastern areas, and the spring thaw moves rapidly northward through the Athabasca–Mackenzie district until it arrives at the edge of the Arctic tree line, where it is retarded. The moderating effects of large lakes in the basin extend the length of the frost-free period in this area and postpone the first heavy winter snow cover. Higher summer precipitation and summer solar radiation, plus generally higher mean hours of bright sunshine, all suggest advantageous circumstances for rapid plant recovery. The reverse is true in the Lakes and Gulf areas, where fewer mean hours of bright sunshine, less general summer solar radiation, and long-lasting winter snow cover might delay forage restoration.

Once plant recovery in secondary habitats has advanced sufficiently, herbivore populations could expand into them, and migratory terrestrial predators would be attracted to these areas and might even increase their reproductive rates to take advantage of available prey. As suitable forage in these areas is reduced, herbivores would withdraw into more protected areas, and predators, having exhausted accessible prey, would move to neighboring areas or to whatever distant areas would offer food supplies. When excess populations reach habitat limits, predators (or prey) presumably perish until reduced to numbers commensurate with local carrying capacity.

The concept of an epicenter would seem to be comparable to dispersal from refugia, only on a much larger scale. In boreal Canada this concept is applied to migrating predators tracking prey. In Norway the epicenter concept may be directly applicable to the herbivorous lemming populations. Collett (1895) recorded that five great groups of mountains in Norway determined the radiating centers from which most of the lemming migrations emanated (fig. 75): (1) the mountain plateaux of the Jotunheimen and Lang-Fjeld, sending their swarms downward to the western portion of Christiania (Oslo) County, Kristiansund County, and the adjoining parts of Bergen County; (2) the plateau of the Dovre Fjeld, extending into the Gudbrandsdal, Osterdal, and Romsdal mountains, sending swarms into the northern parts of Christiania County and adjoining portions of Trondheim Stift as far north as Trondheim Fjord;

Figure 75. Centers of lemming migratory populations in Norway: 1, mountain plateaux of the Jotunheimen and Lang-Fjeld; 2, plateau of the Dovre Fjeld; 3, frontier mountains of Nord Trøndelag; 4, mountains of Nordland; 5, mountain plateaux of Finnmark and neighboring large islands.

(3) the frontier mountains of Nord Trøndelag, sending hordes westward over the entire county and eastward into Swedish Jemtland, down to the Gulf of Bothnia—here migrations occur more frequently than elsewhere; (4) the mountains of Nordland, from which migrations reach the coast and islands in the west, and spread into Swedish Lapmarken in the east; and (5) the extensive mountain plateaux of Finnmark, and the neighboring large islands, from where the valley and lowlands of Finnmark are invaded in all directions. Areas 1 and 2 are included in Elton's (1942) subdivision Southern Norway, areas 3 and 4 his subdivision Central Norway, and area 5, North Norway.

An examination of these Norwegian fox data, which include both colored and arctic foxes, for each Norwegian province reveals a clearer

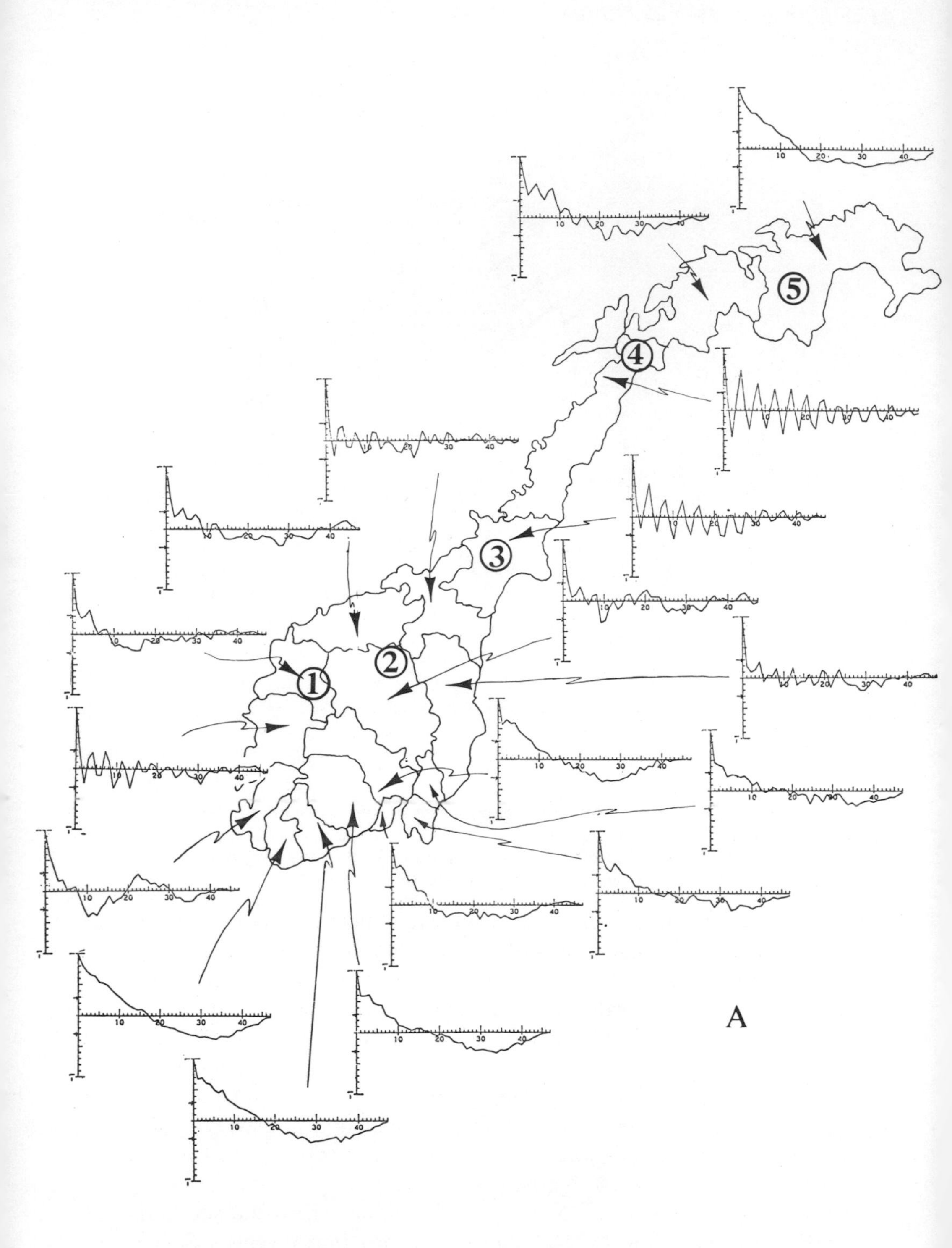

1
2
3
4
5
A

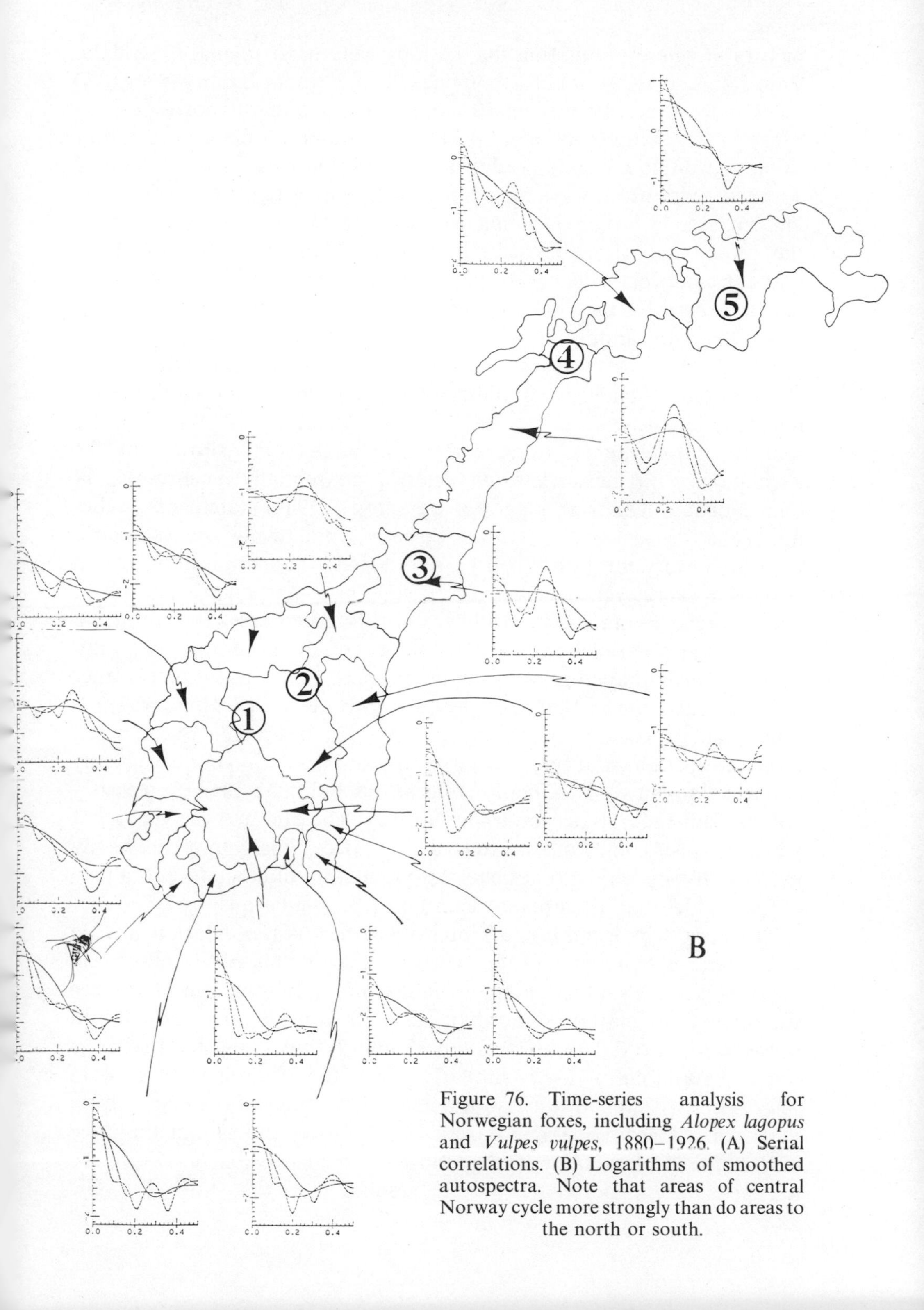

Figure 76. Time-series analysis for Norwegian foxes, including *Alopex lagopus* and *Vulpes vulpes*, 1880–1926. (A) Serial correlations. (B) Logarithms of smoothed autospectra. Note that areas of central Norway cycle more strongly than do areas to the north or south.

picture of where population fluctuations were most regular. The data, from Johnsen (1929), are based on pelts brought in for bounty from 1880 to 1926; the series should each be moved back one year to correspond to Elton's dates. Although the political boundaries are somewhat arbitrary as far as wildlife are concerned, it is obvious that cycles with periods of about 4 years are always associated with one of these five epicenters (fig. 76). Second, it is obvious that areas associated with centers 3 and 4 are the most regular, with oscillating, slowly damping autocorrelation functions and relatively sharp spectral peaks at $f = 0.25$. Since area 3 is said to experience the most frequent migrations, it is not surprising that this area is outstanding. It also seems reasonable to assume that a large part of the clarity of cycles related to centers 3 and 4 is due to the fact that the political boundaries circumstantially coincide fairly closely with migration areas.

It is relatively easy, using correlation analysis, to show that the oscillations in fox furs in various regions are essentially synchronous all over Norway. This was suggested by Elton's (1942) statements (italics mine) that "*somewhere in Norway* there has practically always been a migration every third or fourth year" but that "these migrations do not happen at every cycle peak in every area." This is particularly important because it means that five basically disjoint epicenters are mass-producing lemmings at a parallel pace. However, "essentially synchronous" is not meant to mean that all peaks occur everywhere at the same time—rather that any variation in timing of peaks and troughs *over a long period* is obliterated by a *general* synchrony among areas. In Canada, over a much larger area, the idea of epicenters in a 10-year population cycle is usually connected to outward population migrations, but this explanation clearly will not suffice for 4-year cycles in Norway. The question of synchrony, which suggests some kind of external factor occasionally (not necessarily periodically) knocking things back into phase, should therefore be a part of any model that attempts to describe these oscillations.

There is a clear decrease in the intensity of the 4-year oscillation moving southward in Norway, and the spectral peak for this period effectively disappears in the southern portion of the country. In some southern areas the oscillations may be somewhat distorted from receiving population pulses from two different epicenters (1 and 2), but the general trend is obvious. Wildhagen (1952) similarly noted that lemming peaks were seldom conspicuous in the south and west. A study of the distribution maps for colored (fig. 21) and arctic (fig. 15) foxes shows that the cycle fades at the southern limit of the arctic fox. It is conceivable that the arctic fox pelts might dominate the fur counts in areas showing cyclic fluctuations, but it seems more likely that both colored and arctic fox are following lemming cycles, and that colored foxes exercise their option to go to secondary food sources during lemming dearths, whereas arctic foxes must contend with their dependence on lemming availability. This returns to the idea of the obligate migratory predator tracking prey cycles and, because of its migratory habits in a wildly patchy habitat, creating the appropriate type of dispersal sink for a population cycle to occur.

There is, of course, an anomaly to be considered: the fact that Finnmark, replete with lemmings and having its own migration epicenter, does not evidence cyclic fluctuations in fur returns. One possibility is that the sudden shift in the median about 1910, combined with somewhat low

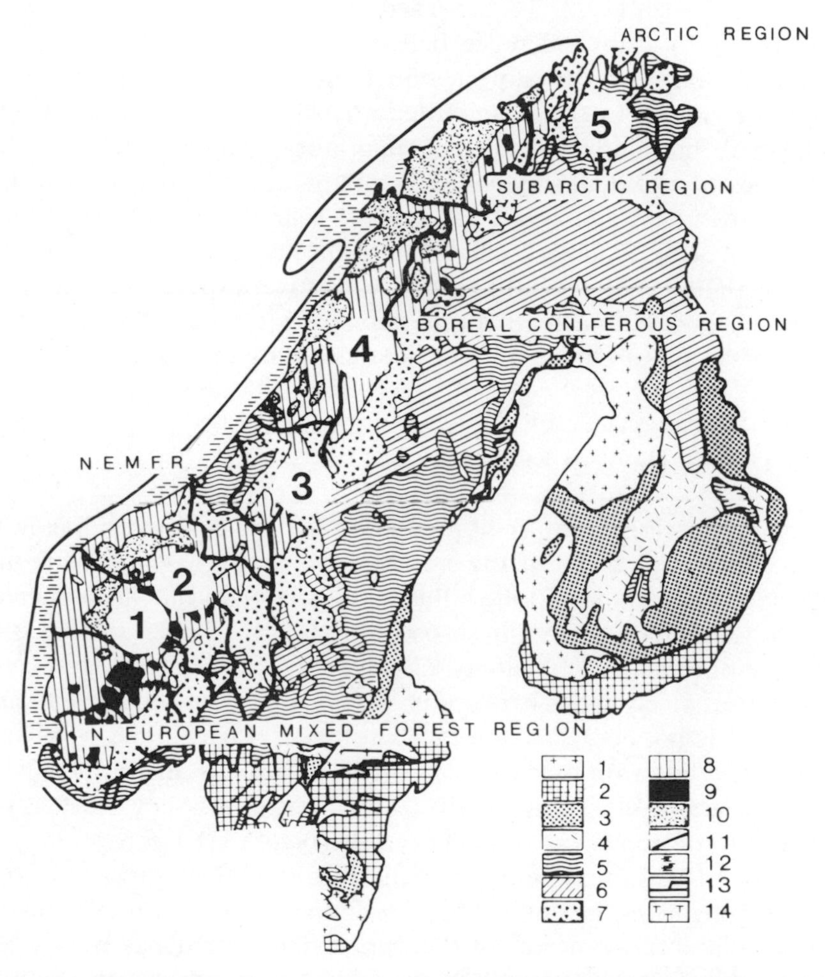

Figure 77. Morphological type regions of Scandinavia. Political boundaries and major centers of dispersal indicated. See fig. 57 for symbolic references. (Based on Rudberg, 1968.)

amplitude cycling, obscures the time-series analysis; however, efforts at smoothing the data by differencing did not improve the situation significantly. If the lemmings in Finnmark evidence migrations, irregular or not, but the predator species (foxes) are not oscillatory, which they appear to be in other regions of Norway, what is the difference?

I would suggest that the difference may be basic *geography*. The other four migration centers are all characterized by fjells with high alpine relief (fig. 77), a type of relief common in western and northern Norway, but elsewhere restricted to the highest mountains (Rudberg, 1968). Finnmark, however, is a low-lying subarctic region covered with low, sparse birch forests over large areas, something uncommon in arctic regions (Hustich, 1968), and Collett (1911–12) recorded that in these arctic parts, where there is only a slight difference between high and low fjells, lemming migrations tended to be shorter, and unhindered individual lemmings could later return to the same plateau from which they began. Collett (1895) had previously noted that migratory lemmings in Finnmark sometimes occupied all the low-lying valleys as far as the sea, and they might settle there and breed for several years after the migration. If dispersal areas in Finnmark offer more protective cover for the migrants, allowing them to survive until reduction of cover exposes them to heavy predation, population fluctuations may have a reduced amplitude. It would be useful to pursue a more detailed study of this area.

What Do We Have to Explain?

To summarize: (1) a number of prey and their predators, residing in the arctic and boreal zones of the northern hemisphere, evidence regular population fluctuations; (2) these fluctuations are basically local phenomena, although data reveal these oscillations on a larger scale; (3) the structure of the habitat ("patchy island") may be important to create a situation in which cycles can occur; (4) predation is a large element in cycles, but is not necessary for cycles to occur; (5) a simple population growth equation with time lags can oscillate, but it is unclear why, since this produces cycles quite easily, more cycles do not exist; (6) an interaction of some nature between herbivore and forage may be important in explaining cycles; (7) a number of mathematical viewpoints have proven interesting but of limited usefulness; (8) no clear evidence of a cyclic extrinsic phenomenon of the appropriate period has been clearly correlated with population cycles in a direct way, but the possibility of some external timing mechanism has not been discounted; (9) stress may be involved in cycles, but more likely as effect than cause; (10) migration and dispersal are major elements in cyclic fluctuations; and (11) an explanation of synchrony should be included in any model of cyclic population fluctuations.

How can these be put together to explain why cycles occur, and why they are relatively regular?

5 *Nati Sunt Mures, et Omnia in Ordine Posita Est*

... so that, after many mice have been born, we may, in ten or twenty years, have a companion volume in which the confusion that now permeates the whole subject will be resolved into order.

G. Evelyn Hutchinson (1942)

When trying to assess causal factors for cyclic fluctuations, it seems useful to avoid the dualistic attitude of separating potential contributing factors into basically opposing camps: the extrinsic camp, emphasizing weather, food shortage and nutrient cycling, predation, and disease; and the intrinsic camp, emphasizing genetics, behavior, and stress. It is no recent revelation that ecological systems are complex, and that the linear-causality approach (*this* causes *that*) seldom leads to a reasonably accurate description of what we observe. It seems crucial to emphasize "interconnection, wholeness, qualitative relations, multiple causality, the unity of structure and process, and the frequently contra-intuitive results of contradictory processes" (Levins, 1974). Many ecologists who, like myself, persistently pursue underlying principles in ecology but, unlike myself, are somewhat skeptical of more sophisticated mathematics, often feel that this is like the gate to Dante's Inferno: "Abandon hope, all ye who enter here." I firmly believe that ecological experience suggests quite the contrary, and that our skepticism is based on the frequent failure of mathematical ecology to convince the naturalist that, having suffered trial by equation, he has been led to a new ecological insight, or been given a new direction for research. This difficulty seems to be in part a consequence of theoretical ecologists being frequently more interested in the mathematics of complexity per se than in ecological understanding.

Of course, this is partly historical circumstance. When we first approach a system, we usually see it as a black box and begin by trying to describe its behavior in terms of statistically significant variation in whatever variables we are able to measure. This is something of an input/output approach, which can later help in testing hypotheses concerning the structure of the system (revealed by other approaches), or can allow predictions concerning the behavior of similar systems. However, by itself this approach cannot lead to an understanding of the contents of the black box (Levins, 1974). Thus after one recovers from the initial excitement of recognizing that spectral analysis clearly demonstrates that we have to contend with the reality of cycles in some populations, we are faced with trying to explain why this occurs, and we are simultaneously faced with the realization that spectral analysis is not the answer to this question, and neither are the various mathematical models, such as the Lotka – Volterra equations, that may describe the observed output by specifying an unsupportably simplified view of the contents of the system. Similarly, when we are able to describe the dynamics of a system by a simple equation, we have to ask what, biologically, all the variables specifically represent. If this simple equation is general even within the narrow confines of a species, we have to ask why, for example, foxes appear to oscillate in certain parts of Norway but not in others, or why snowshoe hares oscillate in the Canadian boreal forest but not in Colorado (Dolbeer and Clark, 1975), or rather, perhaps, why oscillations are conspicuous in some areas but inconspicuous, or nonexistent, in others.

Loop Analysis

One way to approach this problem is to try to identify what appear to be the most important variables in the system and to specify, as much as possible, a deterministic theory connecting these variables. This theorem can then be simulated on a computer to see if the observed input/output relations are maintained. However, most ecological systems have a large number of links, many of which may be difficult to measure; a large number of variables, some of which may be vaguely defined; and sometimes a number of partially specified general rules which define a region of possible behavior rather than one specific behavior. In fact, it seems likely that in most cases our total knowledge of any specific system, even the simplest, may always be relatively meager.

There is, however, a statistical method of interpretation that is intermediate between a fully specified deterministic explanation and a listing of statistical observations or identification of groups of related variables. This is *path analysis*, an approach developed in biology by Sewall Wright (1921, 1960, 1978) for examining relationships among several variables. Because of usually inadequate data, this method has not been applied to ecological systems, but the general ideas can be quite useful for qualitatively analyzing a partially specified system, the type of system most frequently encountered in ecological studies. For this reason Wright's path analysis has evolved into Richard Levin's *loop analysis* (Levins, 1974, 1975). Path analysis, and hence loop analysis, are mathematically comparable to input/output economics applied to energy flow through ecosystems. The analysis of the diagrams (directed or signed graphs, or diagraphs) employed is becoming increasingly extensive, and therefore the usefulness of this approach is expanding (e.g., consider Glass, 1975; Roberts, 1976).

Often a signed diagraph, or loop diagram, represents the most detailed mathematical description of a complex ecological system available. As Maynard-Smith (1974) observed, in order to use a mathematical model, it is not necessary for the model to describe precisely the ecological situation, since we are basically observing changes in model behavior corresponding to changes in parameters, and seeing whether similar changes are observed in the ecological system. Of importance is the stability of the system's behavior with respect to variations in parameters: we are usually looking for a system that exhibits a stable pattern of behavior while allowing for numerical inconstancy in all the state variables. In the case of cycles, we are seeking models that can produce periodic fluctuations over a narrow frequency range without necessitating constant parameters and without risking extinction of the species.

Even with a highly simplified general description, precise conclusions can often be reached especially concerning such questions as the stability of a system and its tendency to oscillate. Qualitative analysis can also indicate when variables can be grouped, or when more variables need to be added. These conclusions follow simply from the structure of the community as represented by the loop diagram—which means, of course, that some biologists' qualitative decisions (theoretical interpretations) are built into the network (Levins, 1974).

It is unlikely that the importance of viewing ecological systems as complex networks can be overemphasized. An ecologist begins designing a model based on what he or she has observed as being of obvious importance in the system. As more observations are made, new parameters and pathways are added, but the effects of a change in, or addition of, a pathway may feed back on the initial variables in ways quite different from the initial one-step links. Thus the original predicted effects of one variable on another may be drastically altered by adding or changing (in direction or magnitude) interactions anywhere in the network (Lane and Levins, 1977).

Following Levins (1975), we construct a diagram in which each point or vertex represents a systemic variable. A variable can be any time-dependent factor related to the system and, significantly, can even include variances in a variable, covariances between variables, averages, and other measures (Levins, 1978). These vertices are then connected by lines that indicate the qualitative interactions between the two variables. Graphically, a variable X_1 may augment the rate of change of another variable X_2 (arrow):

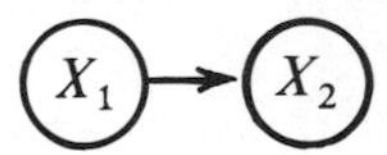

or X_1 may inhibit the rate of change of X_2 (circle):

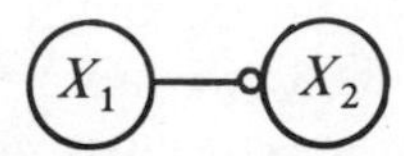

or a variable may have a positive or negative effect on itself:

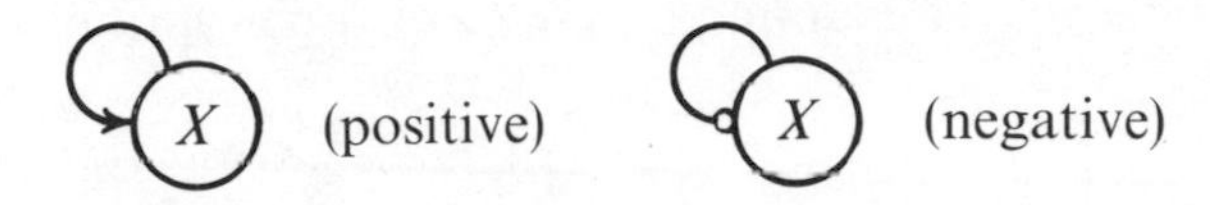

Differential equations, describing rates of change for each variable as a function of all n variables, express the nature of the links among the variables X_i:

$$\frac{dX_i}{dt} = f_i(X_1, X_2, X_3, \ldots, X_n; C_1, C_2, \ldots)$$

where the set of parameters C may represent time-independent parameters relating to the environment (temperature, rainfall averages, etc.) or to the biology of the organisms (growth rates, etc.). In the neighborhood of any equilibrium point (a point at which all the dX_i/dt are zero) the system's behavior depends on the properties of what is called a community matrix $\mathbf{A}$:

$$\mathbf{A} = \begin{bmatrix} a_{11} & a_{12} & a_{13} & \cdot & \cdot & \cdot & a_{1n} \\ a_{21} & a_{22} & a_{23} & \cdot & \cdot & \cdot & a_{2n} \\ a_{31} & \cdot & & & & & \\ \cdot & & \cdot & & & & \\ \cdot & & & \cdot & & & \\ \cdot & & & & \cdot & & \\ a_{n1} & & & & & \cdot & a_{nn} \end{bmatrix}$$

where

$$a_{ij} = \frac{f_i}{X_i}$$

which represents the effect (with respect to direction and magnitude) of a change in variable X_j on variable X_i, and the a_{ij} are the coefficients of the X_j in the equations for dX_i/dt at an equilibrium point. Thus a_{ij} represents the average effect of X_j on the rate of change in X_i. "Average" is stressed because, for example, a herbivore population grazing on a field may actually increase plant productivity in the area while actively destroying individual plants.

Although an actual value for a_{ij} may often be difficult to obtain, the sign of a_{ij} proves to be quite informative, and it is on this fact that the great value of loop analysis depends. For example, if X_j represents a predator species preying on species X_i, a_{ij} is negative and a_{ji} is positive; if X_i and X_j compete, a_{ij} and a_{ji} are both negative. In the diagrams an arrow indicates positive a_{ij} and a circle negative a_{ij}.

Now consider as an example a system of three variables and all its possible links:

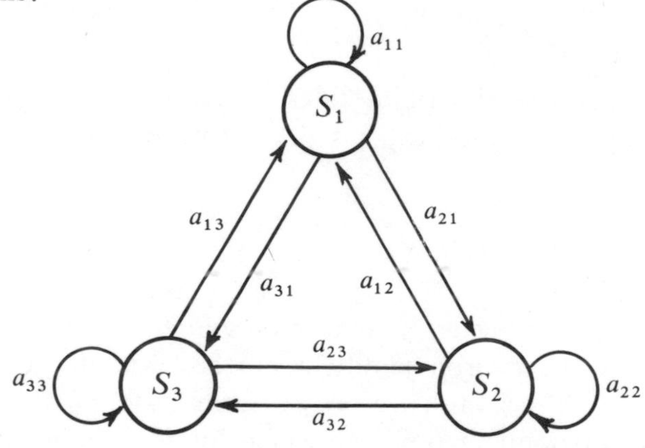

Any link, or series of links, which leaves and eventually reenters a vertex is called a *loop*. A loop of length k is a closed path from any vertex to itself through k links, which passes through each vertex of the loop only once. The value of the loop is the product of the a_{ij} of its links, with the sign determined by the sign of that product. A loop of length zero has by convention a value of $+1$.

The community matrix of this three-variable system would be

$$\mathbf{A} = \begin{bmatrix} a_{11} & a_{12} & a_{13} \\ a_{21} & a_{22} & a_{23} \\ a_{31} & a_{32} & a_{33} \end{bmatrix}$$

The determinant expansion of this matrix is

$$\mathbf{D} = a_{11}a_{22}a_{33} - a_{11}a_{23}a_{32} - a_{12}a_{21}a_{33} - a_{13}a_{22}a_{31} + a_{12}a_{23}a_{31} - a_{13}a_{21}a_{32}$$

which is clearly the product of the three self-loops of length $k = 1$, followed by the products of each self-loop with its corresponding disjunct loop (loop with no variables in common with the first loop) of length $k = 2$, followed by the two loops of $k = 3$ (representing clockwise and counterclockwise passage through the three variables). Each term thus represents the product of loops that go into each variable once and leave it once (i.e., the product of one or more disjunct loops). The sign of an individual term is determined in the usual way by taking the principal diagonal $a_{11}a_{22}a_{33}$ as positive and determining the number of permutations of second subscripts necessary to generate the term, changing the sign once for each permutation (Hutchinson, 1978). Thus an abbreviated representation of the determinant $\mathbf{D}$ would be

$$\mathbf{D}_n = \sum (-1)^{n-m} L(m, n)$$

where $L(m, n)$ is the product of n links a_{ij} of m disjunct loops.

With this terminology it is possible to define feedback, F_k, at level k in the system to be the sum of all possible products of disjunct loops involving k vertices:

$$F_k = \sum (-1)^{m+1} L(m, k)$$

The sign is adjusted to make F_k negative if all loops are negative. If m, the number of disjunct loops in each term of the feedback equation, is odd, feedback will be positive; if even, negative.

Returning to the three-vertex example, the feedback at level $k = 1$ will be

$$F_1 = \sum_i a_{ii} = a_{11} + a_{22} + a_{33}$$

all terms being positive since each self-loop involves only one vertex ($m = 1$). At level $k = 2$,

$$F_2 = \sum_{ij} a_{ij}a_{ji} - \sum_{ij} a_{ii}a_{jj}$$
$$= a_{12}a_{21} + a_{23}a_{32} + a_{13}a_{31} - a_{11}a_{22} - a_{11}a_{33} - a_{22}a_{33}$$

The first group of terms represents single loops ($m = 1$), and is positive, the second group represents products of two disjunct loops ($m = 2$) and is negative. At level $k = 3$,

$$F_3 = \sum_{ijk} a_{ii}a_{jj}a_{kk} - \sum_{ijk} a_{ii}a_{jk}a_{kj} + \sum_{ijk} a_{ij}a_{jk}a_{ki} = \mathbf{D}_3$$

which is clearly the expansion of the determinant noted above. For algebraic convenience, define $F_0 = -1$.

Given this framework it is possible to address the question most relevant to this study: How does this community system behave? In particular, are there certain conditions which, if met, would produce or allow for oscillations? Levins (1975) offered a means of answering these questions by recalling the fact that the nature of the eigenvalues (λ) of the community matrix describe the stability of the community and that, if an eigenvalue is a complex number, oscillatory behavior is to be expected. If an oscillation exists, its nature will be determined by the real part of the complex eigenvalue: if the real part is negative, the system will be stable to a pulse (i.e., change) at any vertex (see Roberts, 1976) and will return to equilibrium through damped oscillations; if the real part is zero, the system will be neutrally pulse stable and will oscillate with a persistent amplitude dependent on the initial displacement—this would be comparable to the solution of the Lotka–Volterra system of predator–prey equations—and if the real part is positive, the system will be unstable, oscillating about equilibrium with increasing amplitude. If the real part is positive and the system does not oscillate, the perturbed system will move away from equilibrium in the direction of the initial displacement. It is not unreasonable to think of arctic areas as persistently perturbed by such extrinsic factors as weather (consider Fuller et al., 1969; Pitelka, 1973), which might act directly on the fauna or indirectly through variations in herbivore forage.

The characteristic equation $P(\lambda)$ from which the eigenvalues are derived is a polynomial in λ of order equal to the number of variables X_i in the system. This can be shown to be simply related to the expression for feedback:

$$P(\lambda) = \lambda^n - \sum_k F_k \lambda^{n-k}$$

If all the roots of this equation are to have negative real parts, as is required for a stable system that returns to a steady-state value after perturbation, net feedback F_k at each level k must be negative. In addition, excessive negative feedback with sufficiently long time lags (which are not explicit in systems of differential equations but which enter indirectly through the effects a variable has on itself via other variables, i.e., via loops of various lengths) will result in oscillatory instability; to avoid this, negative feedback from long loops must not be too strong compared to negative feedback from short loops. A series of exprcssions, all of which must be positive, define this condition, but for this analysis the concept can be sufficiently described by

$$F_1 F_2 + F_3 > 0$$

It is useful to recall that this discussion concerns the local stability properties of the steady-state equilibrium, since the argument necessitates linearity in the operation of the coefficients a_{ij}. However, the results may offer new directions of thought, and it is encouraging to know that many properties of the global dynamics of biological systems can be similarly deduced by analyzing the qualitative interactions between elements of the system (Glass, 1975).

How can loop analysis be used to clarify population oscillations? Consider a two-variable system, for example a plant–herbivore system or a prey–predator system, and its associated community matrix:

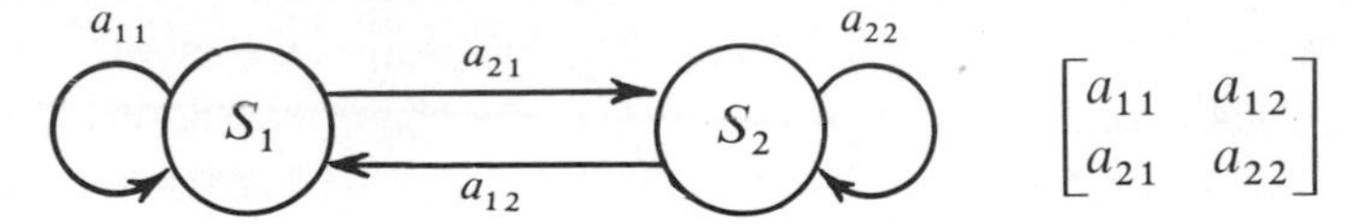

Feedback at levels $k=1$ and $k=2$ would be

$$F_1 = \sum_i (-1)^{1+1} a_{ii} = a_{11} + a_{22}$$

$$F_2 = \sum a_{ij} a_{ji} - \sum a_{ii} a_{jj} = a_{12} a_{21} - a_{11} a_{22}$$

The characteristic equation of this matrix is

$$\begin{aligned} P(\lambda) &= \lambda^2 - (a_{11} + a_{22})\lambda + a_{11} a_{22} - a_{12} a_{21} \\ &= \lambda^2 - (F_1)\lambda - F_2 \end{aligned}$$

with solution (via the quadratic formula)

$$\lambda = \frac{F_1 \pm \sqrt{F_1^2 + 4F_2}}{2}$$

For a stable damped oscillation,

$$\left.\begin{aligned} F_2 < 0 \\ F_1 < 0 \end{aligned}\right\} \quad \text{(for stability)}$$

$$F_1^2 + 4F_2 < 0 \quad \text{(for oscillation)}$$

If it is necessary for stability that

$$F_1 = a_{11} + a_{22} < 0$$

at least one variable requires negative self-feedback. If neither variable has self-feedback,

$$F_1 = 0 \qquad \text{(which in this system means neutral stability)}$$
$$F_2 = a_{12}a_{21} < 0 \qquad \text{(to be oscillatory)}$$

For this to be true, a_{12} and a_{21} would require opposite signs, for example

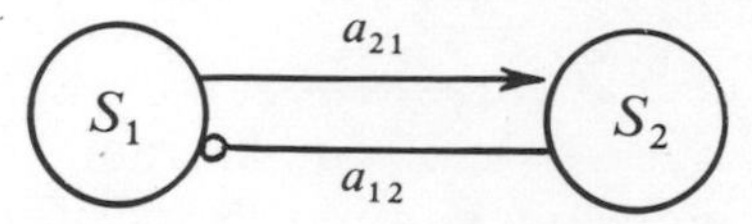

This describes the neutrally stable oscillation of the Lotka–Volterra system. Since a neutrally stable oscillation can experience local extinction, an additional factor (e.g., emigration) would be required for persistence, which means introducing either feedback or another variable, leading to a different system. Once self-feedback is introduced, oscillation depends on actual values of the coefficients and not just on the nature of the interrelationships.

What kind of community structure might produce a stable, oscillatory situation, which is what cycles seem to represent? Consider the factors that seem to have possible roles in cycles: several intrinsic reactions, such as stress, variations in reproduction, and so on, may have feedback effects on the population, either by directly leading to reduced numbers or by stimulating a variation in behavior patterns (e.g., dispersal); external factors may play a part in maintaining a perturbed state in the system either regularly or occasionally; dispersal seems to play an important part, and predation also appears to be a significant factor. Also, the idea of survival habitats and colonizing habitats seems a useful inclusion, and forage limitations or nutrient cycling may be important. Considering these factors, examine the following system, shown with its corresponding community matrix:

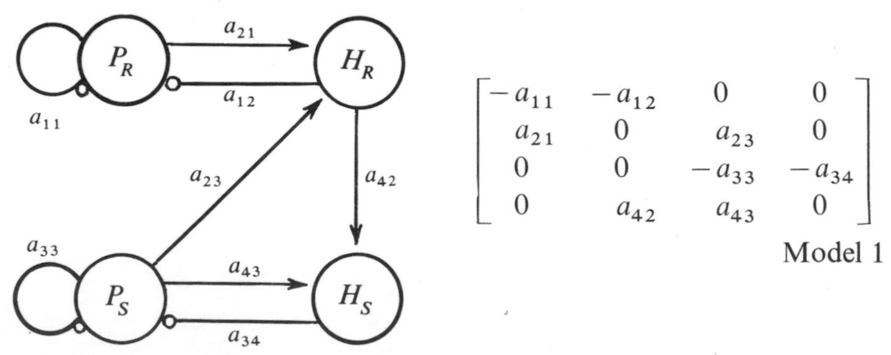

$$\begin{bmatrix} -a_{11} & -a_{12} & 0 & 0 \\ a_{21} & 0 & a_{23} & 0 \\ 0 & 0 & -a_{33} & -a_{34} \\ 0 & a_{42} & a_{43} & 0 \end{bmatrix}$$

Model 1

In this system define a refuge R with a self-limiting plant population P_R and resident herbivore H_R, and define a colonizing habitat, or dispersal sink S with a self-limiting plant population (P_S) and herbivores who have migrated from the refuge and may or may not have become resident (H_S). An increase in herbivores in the refuge stimulates emigration to secondary habitats, or the dispersal sink, causing an increase in the herbivore population in the sink ($+a_{42}$). An increase in forage in the dispersal area signals the refuge herbivores that emigration has survival value ($+a_{23}$), and stimulates population growth there. Feedback at all levels is negative, indicating that the system is stable (i.e., any perturbation will be followed by a return to equilibrium):

$$F_1 = -a_{11} - a_{33} < 0$$
$$F_2 = -a_{21}a_{12} - a_{43}a_{34} - a_{11}a_{33} < 0$$
$$F_3 = -a_{23}a_{42}a_{34} - a_{33}a_{21}a_{12} - a_{11}a_{43}a_{34} < 0$$
$$F_4 = -a_{11}a_{34}a_{23}a_{42} - a_{21}a_{12}a_{43}a_{34} < 0$$

Solving for the oscillatory criterion, we obtain

$$F_1F_2 + F_3 = a_{11}a_{21}a_{12} + a_{33}a_{43}a_{34} + a_{11}^2a_{33} + a_{11}a_{33}^2 - a_{23}a_{42}a_{34}$$

For the system to be oscillatory, this expression must be less than zero, which means that the term ($-a_{23}a_{42}a_{34}$) must be quite large relative to the other positive terms. This would be true if emigration is a major impulse [since all three of the factors in this term relate to emigration: stimulation of emigration by available forage in dispersal areas (a_{23}); increase in colonizing population due to emigration (a_{42}); and effects of emigrants on forage in dispersal areas (a_{34})] and if self-inhibition by plants (a_{11} and a_{33}) is of relatively minor import. Observations on both hares and lemmings suggest that dispersal from a refuge is, in fact, important (recall Tamarin, 1978, or Wolff, 1977a, for example) and, because of the structure of the habitat, possible, and it does not seem too presumptuous to suggest that grazing herbivores may keep plant populations below the level of sizable negative self-feedback. Thus this system seems consistent with the observed structure of cyclic communities.

But what happens when carnivorous predation (K) is included in this system? Consider the system

$$\begin{bmatrix} a_{11} & -a_{12} & 0 & 0 & 0 \\ a_{21} & 0 & a_{23} & 0 & 0 \\ 0 & 0 & -a_{33} & -a_{34} & 0 \\ 0 & a_{42} & a_{43} & 0 & -a_{45} \\ 0 & 0 & 0 & a_{54} & 0 \end{bmatrix}$$

Model 2

Feedback at all levels is, again, negative:

$$F_1 = -a_{11} - a_{33} < 0$$
$$F_2 = -a_{21}a_{12} - a_{43}a_{34} - a_{45}a_{54} - a_{11}a_{33} < 0$$
$$F_3 = -a_{23}a_{42}a_{34} - a_{11}(a_{43}a_{34} + a_{45}a_{54}) - a_{33}(a_{21}a_{12} + a_{45}a_{54}) < 0$$
$$F_4 = -a_{11}a_{23}a_{42}a_{34} - a_{21}a_{12}(a_{43}a_{34} + a_{45}a_{54}) < 0$$
$$F_5 = -a_{33}a_{21}a_{12}a_{45}a_{54} < 0$$

Similarly,

$$F_1F_2 + F_3 = a_{11}a_{21}a_{12} + a_{33}a_{34}a_{43} + a_{11}^2a_{33} + a_{11}a_{33}^2 - a_{23}a_{42}a_{34}$$

which is precisely the same expression as in the absence of predators. Therefore, *those same factors that would produce oscillatory instability in the absence of predators would produce it in their presence.* Also, migratory predators, usually associated with population cycles (especially in far northern climates), would not be expected to affect systemic stability either because of their presence or absence. This result supports the ideas that cycles may exist without predation (presuming the habitat structure is as defined here); that the role of the migratory obligate predators may be to force the migrant herbivore population to an extreme low, which would enhance the extremes of amplitude of the cycles; and that the population decline can begin before predation becomes significant. The fact that the tundra predators tend to be obligate and migratory can be appropriately reemphasized since the absence of prey in dispersal zones, and the relative inaccessibility of prey in refuges, force the predators to go elsewhere. The absence of the predator allows dispersal areas to be actually colonized rather than having the prey continually removed at the refuge perimeter, a situation soon to be examined. However, with or without the predator, oscillatory instability is possible in the herbivore population, and the role of the predator may be simply enhancement of cyclic amplitudes.

It is reasonable to expect that should a migratory predator arrive while vegetation in dispersal zones is dense, providing reasonable cover for the herbivore population, the predator's activity will be negatively affected. This would be described by

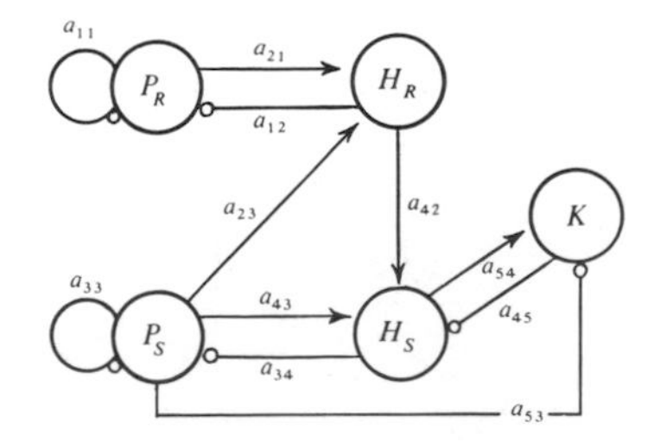

$$\begin{bmatrix} -a_{11} & -a_{21} & 0 & 0 & 0 \\ a_{21} & 0 & a_{23} & 0 & 0 \\ 0 & 0 & -a_{33} & -a_{34} & 0 \\ 0 & a_{42} & a_{43} & 0 & -a_{45} \\ 0 & 0 & -a_{53} & a_{54} & 0 \end{bmatrix} \quad \text{Model 3}$$

The effect of this feedback from the plants of the colonizing zone on the predator is to increase the magnitude (in the negative direction) of feedbacks at level $k=3$ and above, plus adding an additional negative term to the oscillatory stability criterion:

$$F_1F_2+F_3=-a_{53}a_{45}a_{34}-a_{23}a_{42}a_{34}+(\text{positive terms})$$

This increases the probability of oscillatory instability in the system.

Returning to the original model, negative feedback could be added to the herbivore populations:

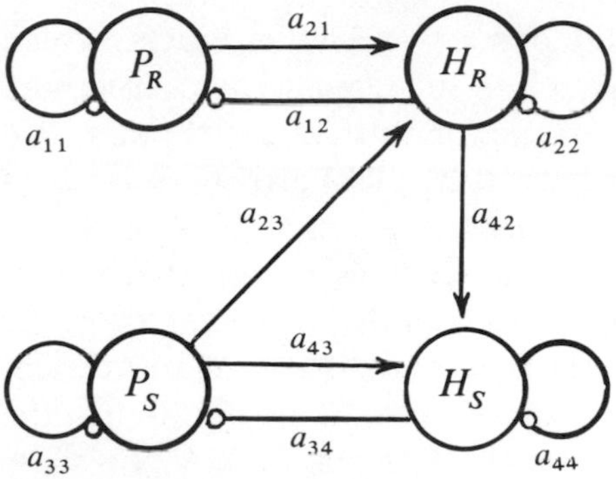

$$\begin{bmatrix} -a_{11} & -a_{12} & 0 & 0 \\ a_{21} & -a_{22} & a_{23} & 0 \\ 0 & 0 & -a_{33} & -a_{34} \\ 0 & a_{42} & a_{43} & -a_{44} \end{bmatrix}$$

Model 4

These negative feedback terms could be interpreted as density-dependent self-inhibition resulting from stress affecting reproduction, maternal behavior, or similar factors (stress might also affect the system by increasing emigration, i.e., increasing a_{42}). An examination of feedback at all levels reveals additional terms in a_{22} and a_{44}, but all feedbacks remain negative:

$$F_1=-a_{11}-a_{22}-a_{33}-a_{44}<0$$

$$F_2=-a_{12}a_{21}-a_{43}a_{34}-a_{11}(a_{22}+a_{33}+a_{44})-a_{22}(a_{33}+a_{44}) -a_{33}a_{44}<0$$

$$F_3=-a_{23}a_{42}a_{34}-a_{11}a_{43}a_{34}-a_{22}a_{43}a_{34}-a_{33}a_{21}a_{12}-a_{44}a_{21}a_{12} -a_{11}a_{22}(a_{33}+a_{44})-a_{11}a_{33}a_{44}-a_{22}a_{33}a_{44}<0$$

$$F_4=-a_{23}a_{42}a_{34}a_{11}-a_{11}a_{22}a_{34}-a_{33}a_{44}a_{21}a_{12}-a_{11}a_{22}a_{33}a_{44}<0$$

Checking for oscillatory instability:

$$F_1F_2 + F_3 = (\text{positive terms involving self-inhibition}) - a_{23}a_{42}a_{34}$$

Again the strength of the "emigration loop" determines the potential for oscillation.

What would be the effect of including a nutrient cycle in this system? The answer can be approached initially by examining a nutrient cycle in isolation:

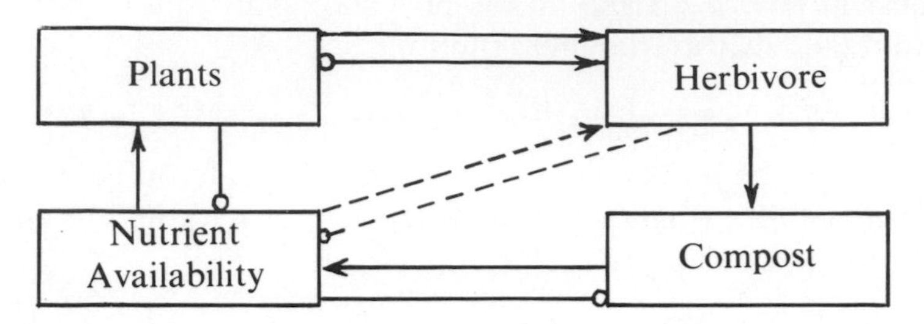

The herbivore consumes plants and deposits waste products, which decompose into nutrients that are taken into the plants. The dashed lines are to suggest that there are two basic types of nutrient cycle: one where the available nutrient is involved in stimulating plant growth (solid lines between nutrients and plants), and one where the availability of a nutrient directly affects the herbivore population (dashed lines) such that, while the nutrient is available through the plants, it is not directly involved in the plant growth process. What is important is that either type of nutrient cycle produces a long positive loop, and this tends to destabilize the system. Consider:

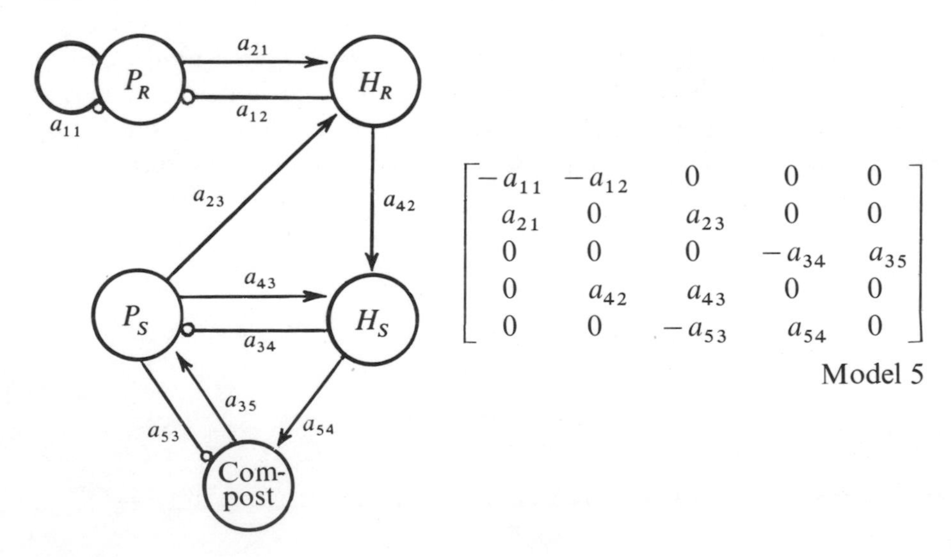

$$\begin{bmatrix} -a_{11} & -a_{12} & 0 & 0 & 0 \\ a_{21} & 0 & a_{23} & 0 & 0 \\ 0 & 0 & 0 & -a_{34} & a_{35} \\ 0 & a_{42} & a_{43} & 0 & 0 \\ 0 & 0 & -a_{53} & a_{54} & 0 \end{bmatrix}$$

Model 5

This describes a nutrient directly affecting the herbivore in the dispersal area. The two long positive loops ($a_{54}a_{35}a_{43}$ and $a_{54}a_{35}a_{23}a_{42}$) suggest possible instability, and it is easily shown that feedback terms at level 3 and higher have positive terms, and in particular that

$$F_5 = a_{11}a_{54}a_{35}a_{23}a_{42} + a_{21}a_{12}a_{54}a_{35}a_{43} > 0$$

so that the system is unstable.

One way in which a nutrient cycle could be included in the system without incorporating a long positive (and hence destabilizing) feedback loop would be to incorporate a physiological state for the herbivore which might represent stored nutrient (S), increased by feeding but decreased by metabolic expenditures (Levins, 1978). An increase in stored nutrient could stimulate herbivore reproduction:

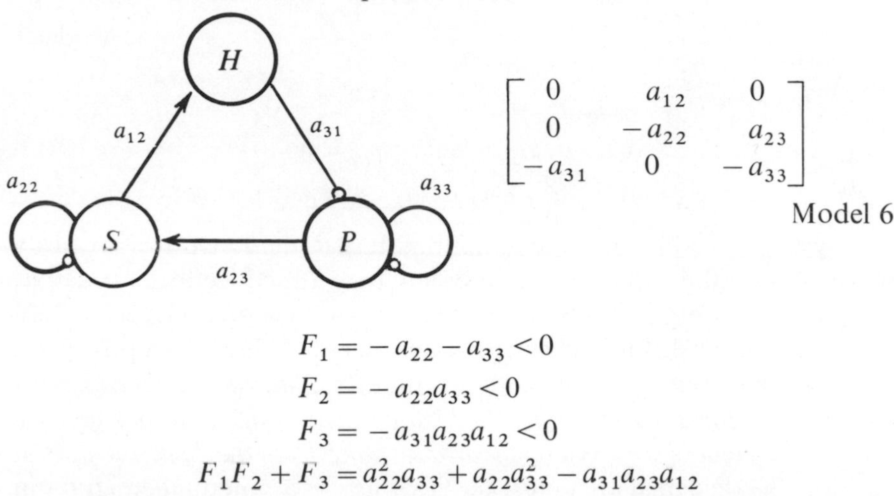

$$\begin{bmatrix} 0 & a_{12} & 0 \\ 0 & -a_{22} & a_{23} \\ -a_{31} & 0 & -a_{33} \end{bmatrix}$$

Model 6

$$F_1 = -a_{22} - a_{33} < 0$$
$$F_2 = -a_{22}a_{33} < 0$$
$$F_3 = -a_{31}a_{23}a_{12} < 0$$
$$F_1F_2 + F_3 = a_{22}^2a_{33} + a_{22}a_{33}^2 - a_{31}a_{23}a_{12}$$

If the link between stored nutrient and increased reproduction (a_{12}) is stronger than self-inhibiting mechanisms, oscillations could result. This outcome would not be modified by adding a predator to the system. However, this model may not be consistent with observations at Barrow, Alaska, on lemming populations, where, with forage in good condition (and, theoretically, nutritious) and predators persistent, the cycle seemed to abate, at least from 1965 to 1973 (Pitelka, 1973). Tamarin's (1978) observations that island voles in Massachusetts did not cycle while nearby mainland voles did cycle may also contradict this simple model. However, the idea of the model may describe a mechanism by which refuge populations could discover that dispersal areas are available for colonization. I suspect that prospective colonists, finding good forage and little competition or predation in a restored dispersal area, may precipitate population increases by simply leaving the refuge and reducing the negative feedback on the refuge population that would arise through impact on refuge forage. This is similar to saying that emigration exerts a type of "positive feedback" on the refuge population by not being negative.

Is the presence of a dispersal sink necessary for this model? For example, could oscillations result as a consequence of interactions between a herbivore and two different plant groups, one in a refuge (protected) area and another in a colonizing (secondary) area, with the herbivore moving from one area to another as forage is available? This could be described as

$$\begin{bmatrix} -a_{11} & -a_{12} & 0 \\ a_{21} & 0 & a_{23} \\ 0 & -a_{32} & -a_{33} \end{bmatrix}$$

Model 7

$$F_1 = -a_{11} - a_{33} < 0$$
$$F_2 = -a_{21}a_{12} - a_{32}a_{23} - a_{11}a_{33} < 0$$
$$F_3 = -a_{11}a_{32}a_{23} - a_{33}a_{21}a_{12} < 0$$
$$F_1F_2 + F_3 = a_{11}a_{21}a_{12} + a_{33}a_{32}a_{23} + a_{11}^2a_{33} + a_{11}a_{33}^2 > 0$$

This system is stable and nonoscillating. It also demonstrates an easily proven fact that in a three-variable system with all self-feedback negative, $F_1F_2 + F_3$ can only be negative if there are negative loops at level $k = 3$. In the same fashion this model system would serve to describe a predator–prey–forage system: the implication is that *the addition of a predator as a dispersal sink for a herbivore in a herbivore–plant system will not produce an oscillation unless there is a refuge area separate from the dispersal area.* In addition, the dispersal area must be "inviting" to a potential emigrant in the refuge area i.e., positive feedback $+a_{23}$ in model 1; if this feedback term is zero, the model will not have a potential for oscillation. However, the distinct separation of a refuge area and dispersal area does not in itself guarantee potential oscillatory instability. Consider the system

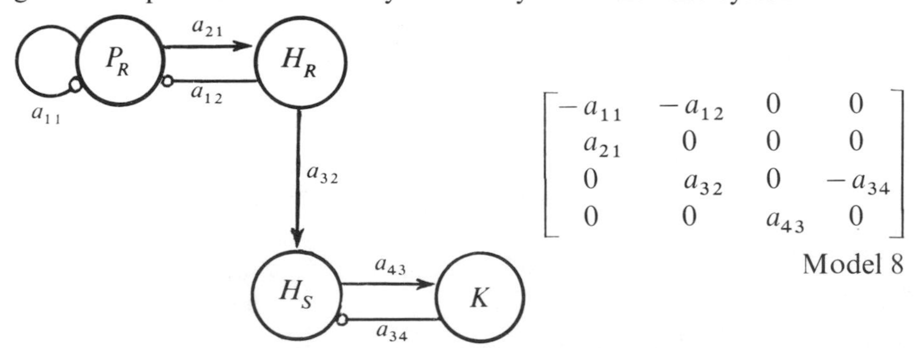

$$\begin{bmatrix} -a_{11} & -a_{12} & 0 & 0 \\ a_{21} & 0 & 0 & 0 \\ 0 & a_{32} & 0 & -a_{34} \\ 0 & 0 & a_{43} & 0 \end{bmatrix}$$

Model 8

which might describe a herbivore in a refuge area supplying excess (or colonizing) population to a predator on the refuge perimeter. For this system,

$$F_1 = -a_{11} < 0$$
$$F_2 = -a_{12}a_{21} - a_{43}a_{34} < 0$$
$$F_3 = -a_{11}a_{43}a_{34} < 0$$
$$F_4 = -a_{21}a_{12}a_{43}a_{34} < 0$$
$$F_1F_2 + F_3 = a_{11}a_{12}a_{21} > 0$$

Therefore, the system is stable and nonoscillatory. This implies that in a heterogeneous tundra ecosystem in which obligate migratory predators allow prolonged dispersion from refuge areas into secondary habitats to occur, the structure of model 1 is most appropriate, and the potential for oscillations exists, whereas in the western United States, where predators are facultative and resident, the system just described is more appropriate and this system would not be expected to cycle. This theory is supported by observations on snowshoe hare in Canada (Wolff, 1977a) and in Colorado (Dolbeer and Clark, 1975). It may be that the differences between cycling lynx and closely related acyclic bobcats could also be summarized in this manner. There is an additional implication that the concept of dispersal sink as a mere drain to remove excess population may be insufficient to produce a population cycle (which was suggested by Tamarin, 1978). Rather, for a system to oscillate there may be a need for a dispersal area that is "inviting" in the sense that it not only encourages emigration from the refuge, but also provides positive feedback to the refuge herbivore population.

This system, with defined refuge and dispersal areas combined with migratory predators, might describe the situation in lemming populations at Barrow, Alaska, from 1965 to 1973, during which period there was no apparent cyclic peak in brown lemming, *L. trimucronatus* (Pitelka, 1973). A late winter storm in June 1965 greatly diminished the impact of avian predators on the lemming population, thus reducing the effect of migrant predation; least weasels (*Mustela rixosa*), another important predator, maintained a continuous predation pressure during the summer and subsequent winter, but the brown lemming population entered the winter with higher-than-normal densities for a postpeak season (peak in 1965). In 1966, brown lemming numbers were low, and mammalian predators may have been present. In 1967–68 weasels abounded and, while lemming numbers increased during the summer of 1968, the number was not as large as during typical prehigh summers; the presence of weasels in numbers was unprecedented. An unusual ice condition during the winter of 1968–69 forced lemmings to forage along runway edges rather than freely under the snow, and an attempted population increase was apparently aborted; exceptional numbers of weasels during the winter and summer kept the population down. A severe winter in 1969–70 continued to keep numbers down; lemmings were scarce and weasels were absent in the summer of 1970. Consequently, vegetation was in superb condition in the summer of 1971, and large numbers of immigrant lemmings appeared. However, arctic foxes, another winter lemming predator, were common in the fall and early winter of 1971, until Eskimo fox trapping began on December 1.

This description clearly suggests that the effects of predation, for various reasons, had switched from a migratory-type effect to a resident-type effect, and cycles, as the system's models suggest, disappeared. Meanwhile, the consequent emergence of lush vegetation provided a functional dispersal area for a neighboring population, with the migrants being heavily preyed upon by the "resident" prey.

The unprecedented appearance of collared lemmings (*Dicrostonyx*) at Barrow in 1970 and 1971 is consistent with this dispersal system concept. *Dicrostonyx* is generally found in well-drained upland habitats and high ridges, which are not common at Barrow, whereas the brown lemming, generally most common in wet lowland habitats, will occur in essentially all habitats during population highs. However, the reduced brown lemming population in 1970 and early 1971, combined with the lush vegetation, offered a viable dispersal area for *Dicrostonyx*. The 1971 influx of migrant *Lemmus*, a more aggressive species than *Dicrostonyx*, shifted the balance back toward *Lemmus*. This again supports the idea of dispersal from a protected refuge into an inviting, and available, dispersal area.

If there is no dispersal area, a simple stable system might evolve:

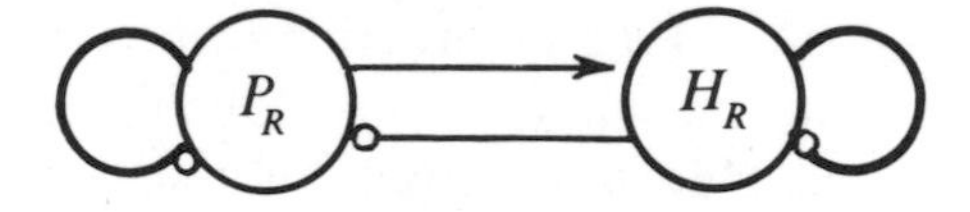

which seems a reasonable model to describe the nonoscillatory vole population, with no significant predation or dispersal sink, on Muskeget Island (Tamarin, 1978). But if the Muskeget system has stabilized in the absence of predation and a dispersal sink, what would account for the violent fluctuations in *M. pennsylvanicus* observed by Krebs et al. (1969)? Returning to the dispersal model and isolating a plant–herbivore group (which basically describes the conditions of the experiment) the system could be represented as

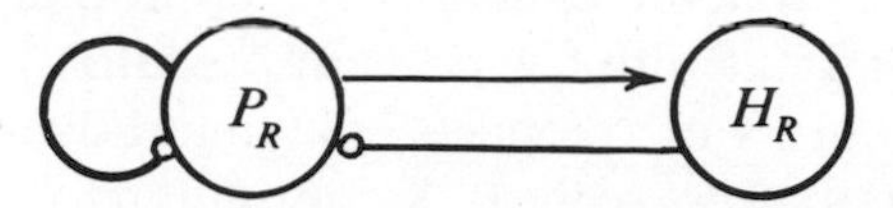

That is, there is no negative feedback in the herbivore population. It has been previously demonstrated that this would yield an oscillatory population if

$$F_1^2 + 4F_2 < 0$$

which in this case means that

$$a_{11}^2 - 4a_{12}a_{21} < 0$$

If self-limitation in the plant is weak relative to plant–herbivore interaction, this would be true. However, the system would be stabilized by the emergence of strong self-limitation in the herbivore population; thus

$$F_1^2 + 4F_2 = (a_{11} + a_{22})^2 - 4(a_{12}a_{21} + a_{11}a_{22})$$

would become positive (and hence the system would be nonoscillatory) if the square of the sum of the self-feedback terms were to increase more rapidly than their product. That self-feedback in a vole population isolated from predators and dispersal possibilities can occur is suggested by the fact that there is more "stress" in the nonoscillatory Muskeget Island vole population than in the cyclic mainland vole population nearby (Tamarin, 1978). This may be the type of stabilization envisioned by Tamarin (1978) when he suggested that Lidicker's (1973) Brooks Island study, in which an introduced population of *M. californicus* demonstrated modified and abbreviated cycles, might be a model outcome of the experiment of Krebs et al. (1969) if the latter had been extended for a longer time period.

Now to return to the original question: Why should selection push these prey systems in the direction of oscillatory behavior? In most predator–prey systems, selection within the prey species for increased viability or fecundity may not affect its abundance (e.g., a whole litter might commonly succumb to predation), whereas selection for more effective predator avoidance offers the possibility of increasing population size (Levins, 1975). A heterogeneous environment offers the possibility for predator avoidance through isolation in protected refuges. A refuge system also offers the opportunity for isolated genetic development within the refuge followed by dispersal into diverse habitats, an ideal opportunity for adaptive genetic changes. A species that is capable of producing high population densities quickly could, under appropriate conditions (i.e., suitable dispersal habitat with few resident predators) inflate numbers in order to release swarms of "pioneers" (Wynne-Edwards, 1962). In this respect Tamarin's (1978) use of the $r-K$ continuum to describe this system seems particularly relevant. K selection would clearly be expected if no inviting dispersal area appeared, and a stable nonoscillating system would result. A viable dispersal area would encourage gene dissemination by stimulating reproduction and emigration (r selection), thus strengthening the negative feedback in the longer loops and shifting the system toward stable oscillatory behavior. The fact that cyclic herbivores have a high reproductive potential and are able to rise to the occasion is therefore probably important. If community structure allows for this kind of cyclic mechanism, selection would reinforce the mechanism in the prey, since prey cycles would prevent predators from increasing to a level where they might destroy the prey species (Leigh, 1975). Note that although this periodic glut of predators might appear to be a predator-escape strategy similar to mast years in conifers (Murphy, 1968), or periodic flowering in bamboo (Janzen, 1976), one cannot directly infer that the details of the mechanisms precipitating these different cycles are similar, or even comparable.

Although it is conceivable that nonlinearities in the operations of some coefficients in the community matrix could occur (e.g., plants probably do not begin to limit their density until a certain critical density has been reached; dispersal-area plants similarly might not stimulate emigration from refuges until a certain density of usable forage has been attained; etc.), the argument from loop analysis seems convincing: community structure probably sets the stage for population cycles. But yet to be explained is what determines the period of oscillation.

Why Are Cycles Relatively Regular?

It seems best to begin this section by clearly stating that I do not purport to have *the answer*. To this I shall quickly add that I am not convinced that the type of questions that have *an* answer are the ones that will produce the most complete understanding of what ecological systems represent. Ecology is for me a study of relationships, and in the present infancy of ecology it seems that any angle of approach that might help to understand the nature of these relationships is worthy of pursuit—this was the initial premise of this volume. What I do have in response to the question of regularity is a collection of strong intellectual inclinations, developed over slightly more than a decade, concerning what these fascinating and perplexing phenomena may represent.

In view of the variability in length of these cycles (Keith, 1974, notes, for example, that the snowshoe hare cycle in Canada varies in length from 8 to 11 years), why has there been for well over half a century such a fuss over "regularity"? The fact is that while the periods are variable, cycles average 10 years in several species over long periods of time, and it seems that this number might somehow indicate the nature of the ecological relationships that create the circumstances in which these oscillations occur. The fact that period lengths near 10 and 4 repeat themselves for widely separated populations over vast areas which are climatologically similar but geographically disconnected make it somehow hard to accept the idea that these numbers are merely circumstantial. And the fact that neighboring populations can be cycling out of phase with the same period length makes this question even more intriguing.

A good conceptual place to begin is Wolff's (1977a) diagram of Keith's (1974) hypothesis concerning cycles in snowshoe hares (fig. 78). Biological information and community structure (as viewed through loop analysis) suggest that the role of predators is to create the type of functional dispersal sink that sets the stage for the cyclic process. Does this mean that the plants are the prime factor determining period length? An affirmative answer is suggested by a recent review article by May (1977) concerning the existence of thresholds and breakpoints in ecosystems that can attain multiple stable states. Consider a three-level trophic system composed of a prey population [$N(t)$] whose dynamical behavior is dependent on the predator level (P) and the vegetational resource level S (which determines

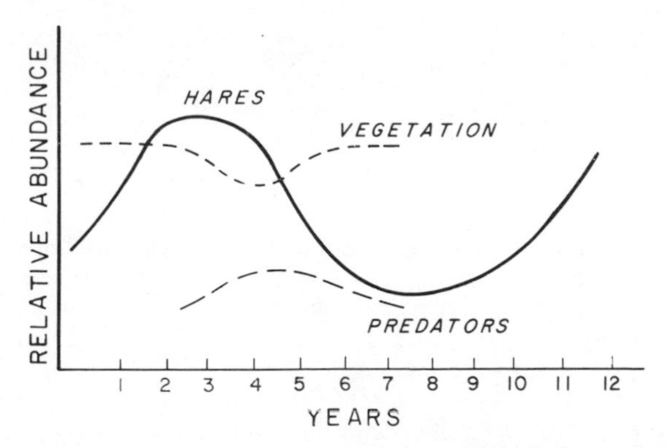

Figure 78. Conceptual model for the 10-year cycle. (From Wolff, 1977a.)

K, the maximum prey level for a given resource level in the absence of predation). If $G(N)$ is the prey growth rate in the absence of predation and $C(N)$ is the consumption rate by the predator population at given N [corresponding to a per capita consumption rate $c(N)$; thus $C(N) = Pc(N)$], the rate of change of the prey population is given by

$$\frac{dN}{dt} = G(N) - C(N)$$

That is, the prey population will tend toward an equilibrium ($dN/dt = 0$) where the natural rate of prey increase exactly balances the loss due to predation. If, as a representative example, the growth rate is taken as logistic and the per capita consumption function $c(N)$ describes predators whose foraging efficiency increases faster than linearly with N at low values of N but saturates to a constant for N above some characteristic value N_0 (Holling's "Type III" vertebrate predator consumption function; Holling, 1965), the result is

$$G(N) = rN\left(1 - \frac{N}{K}\right)$$

$$C(N) = \frac{\beta P N^2}{N_0^2 + N^2}$$

where r is the characteristic prey growth rate and β the saturation attack

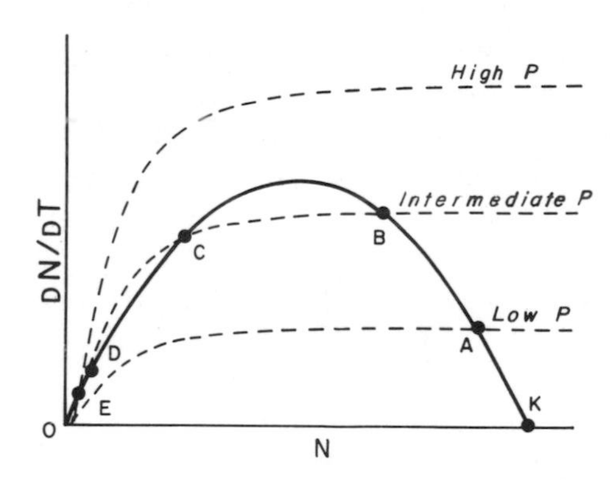

Figure 79. Rate of change of the prey population versus population size for a prey population with logistic growth and a predator population whose foraging efficiency increases faster than linearly at low prey availability but saturates to a constant for large prey numbers (see the text for details), with P, the predator level, varying. Alternative stable states (points B and C) are possible for intermediate levels of P if the per capita predator attack rate saturates to a constant value for values of the predator population significantly less than the prey equilibrium value in the absence of predation. (From May, 1977.)

rate of predators at N_0; thus

$$\frac{dN}{dt} = rN\left(1 - \frac{N}{K}\right) - \frac{\beta PN^2}{N_0^2 + N^2}$$

Figure 79 illustrates this situation for varying P. The essential requirement for yielding alternative stable states for intermediate values of P is that the per capita predator attack rate $c(N)$ saturates to a constant for values of N significantly less than K, the equilibrium value in the absence of predation; that is, N_0/K is significantly less than unity.

For a fixed resource level S, the carrying capacity may be defined as $K = \kappa S$. N_0 is the prey population at which the predator attack rate saturates to the constant level β; this value of N_0 may depend on the prey density relative to the resource level S, since this ratio may determine the nutritive value of a single prey (this may account for varying amplitudes between cycles, since the ratio may not be accurately represented by prey density alone): this may be written $N_0 = \eta S$. Substituting $X = N/\kappa S$, $\tau = rt$, $\gamma = \beta P/r\kappa S$ yields

$$\frac{dX}{dt} = X(1 - X) - \frac{\gamma X^2}{\alpha^2 + X^2}$$

where $\alpha = \eta/\kappa$ represents, for a given value of S, the ratio between the prey density that saturates the predator attack capacity and the maximum prey density that resource level S can sustain. For multiple stable states it is again essential that this ratio be significantly less than unity, and this assumption seems reasonably well supported, at least for snowshoe hares (Keith, 1974; Wolff, 1977a). Once α is fixed, equilibrium value (s) of N/S will depend only on $\gamma = \beta P/r\kappa S$; thus few predators or high resource availability would allow for accelerated prey population expansion.

If the predator level P is fixed at an intermediate level and the equilibrium value(s) of N ($dn/dt = 0$) are plotted against S, an interesting circumstance is revealed (fig. 80). For low values of S, a unique low prey density occurs and the system is essentially under predator control. At T_1, a second stable state appears discontinuously, and for $T_1 > S > T_2$, N will move toward one or the other stable state, depending on which side of the dashed line (called a "breakpoint" locus) it originates: for initial values lying above the dashed line, N will move to the upper equilibrium, and for initial values below the dashed line, N will move to the lower equilibrium. For $S > T_2$, a stable state, relatively independent of predation, exists.

Now, add a simple mathematical form for the rate of change of resource level S, namely a logistic growth term with intrinsic growth rate ρ,

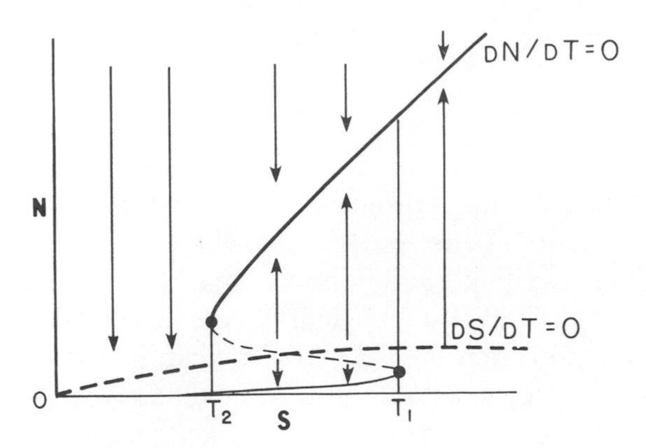

Figure 80. Equilibrium values of the prey population N versus resource level S for predator level P fixed at an intermediate level. For low S, the system is essentially under predator control. But at T_1, a second stable state appears discontinuously. The dashed line between T_1 and T_2 is a breakpoint locus: for initial values above this locus, N will tend toward T_2, and for initial values below the locus, N will tend toward T_1. For $S > T_2$, a stable state, relatively independent of predation, exists. (After May, 1977.)

diminished by losses linearly related to prey density:

$$\frac{dS}{dT} = \rho S\left(1 - \frac{S}{S_{\max}}\right) - \varepsilon N$$

The equilibrium values ($dS/dt = 0$) are sketched in fig. 80.

What kind of behavior might this interactive prey–forage system generate? May notes that this question can be approached by observing that the *time scales* for changes in N and S are quite different. Prey (either lemmings or hares) can reproduce rapidly in response to available forage, but restoration of decimated forage is a lengthy process. Reexamine fig. 80, assuming that prey populations along the lower "predator-controlled" equilibrium curve do not significantly affect the forage and allowing for a slow natural growth in S in this range. Assume further that large prey populations along the upper equilibrium curve have a strongly adverse effect on S. Thus for low values of N and S, the system will move along the fast time scale to the lower prey equilibrium curve along which S can slowly increase, with N quickly adjusting to increasing S. Once the system has reached point T_2, however, continuous change is no longer possible, and the prey population jumps along the fast time scale to the upper equilibrium curve. The adverse effect of prey on forage forces the system to the left along the upper equilibrium curve until, at T_1, N returns along the fast time scale to the lower equilibrium branch.

The result is a cycle of prey "explosions" and "crashes" (fig. 81). Since most of the cycle is spent at low prey densities on the lower equilibrium

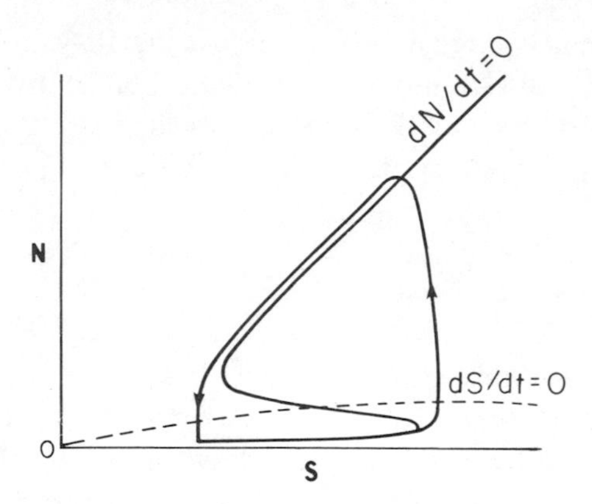

Figure 81. Resource level *S* related to prey density *N* when the rate of change of the resource level increases logistically but is diminished by losses linearly related to prey density. In this situation prey density can increase rapidly in response to available forage, but forage recovery after decimation is slow. See the text for details. (From May, 1977.)

curve, the cycle length is determined by the slow time scale i.e., by the time scale of forage recovery). This time period may therefore define the essential difference between 4-year and 10-year cycles.

Thus the presence of an effective dispersal sink, probably created by the presence of migratory predators, appears to set the stage for oscillations. The resultant level of predation, at which the prey density that saturates the predator attack capacity in the dispersal area becomes significantly less than the maximum prey density that can be supported ($\alpha \ll 1$), may allow for multiple stable states, which can make the system inherently cyclic, the cycle length being determined basically by the forage recovery period. Migrant predators may force the prey population to an extreme low, but the fast response time of the prey species to forage availability will probably obscure any effect predation might have on the length of the cycle.

If it is the structure of the system, including geography, prey–predator relationships, forage recovery, and so on, that produces these oscillations, it may be this same structure that prevents natural selection from eliminating this seemingly costly fluctuation and evolving a more stable system. This is implied in a study by van Zyll de Jong (1975). Noting a relative uniformity in species of Canadian lynx (*L. canadensis*) in marked contrast to extensive subspeciation in the bobcat (*L. rufa*) for a comparable area, van Zyll de Jong suggested that the weak subspeciation in the lynx might be attributable to a uniform distributional range (the boreal forest) and a single primary prey species (snowshoe hare), producing a somewhat uniform selective pressure throughout the range of the lynx. Additionally, extensive dispersal during 10-year peaks, primarily among yearlings, could promote gene flow and offset any local variations that might arise. In this way the structure of the system may determine a situation in which a cycle could both originate and persist.

Deus ex Machina

It is inevitable in any discussion of cycles that some driving force, some causal factor, will be sought to explain why periods are relatively regular and why they are not exactly regular. Because cycles of similar periods occur over such vast areas, meteorological effects are often sought. Sometimes these proposed effects take the form of forcing functions of similar period length, such as the abandoned theory attempting to correlate sunspots and 10-year cycles. Sometimes the theories suggest a massive shock, such as volcanic eruptions (Elton, 1924), which knocks cycles of similar periods and shifting phases back into line. Sometimes theories suggest the periodic occurrence of a succession of events, such as a succession of two or more mild winters followed by a return to normal conditions (Keith, 1974).

All of these approaches seek a function with a period length similar to that of the observed cycle. As an alternative point of view, one could consider the possibility of a shorter-term, regular pulse as a cuing mechanism, and connect this to a statement made by Butler (1953): "The speed and direction of the response to the causal factor depends on the innate qualities of each species and the state of its population at that particular time." Consider, only as an example, the conjunction of Venus with the Earth approximately each 1.6 years, and construct a time line with years and "pulses" for each conjunction (fig. 82). Now say, for example, that forage recovery after decimation by snowshoe hares requires about 8 years and assume that a population peak will occur the summer after the first "pulse" directly following forage recovery. For the time line, a peak would occur at years 10, 20, 29, 39, and 48: average over five cycles of 9.6 years, the cycle length suggested by Elton and Nicholson (1942b) for the lynx cycle in Canada. If forage recovery were accelerated in more southerly climates and required, say, 2 years, peaks would be expected at years 4, 7, 10, 13, and 16: average period 3.2 years, the mean between-peak time calculated by Hutchinson (1975) for Andersen's (1957) acyclic hare data (*Lepus europaeus*) in Denmark. Now assume that forage recovery after decimation by lemming populations requires 3 years—a peak would occur every 4 years, as observed. If vegetation recovery after

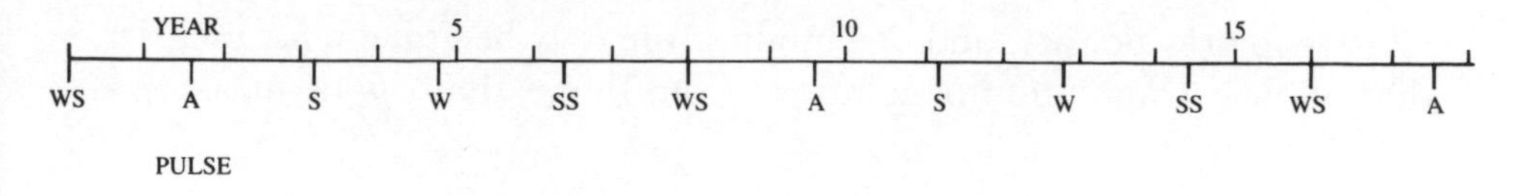

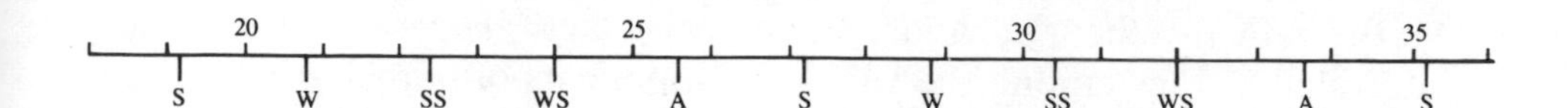

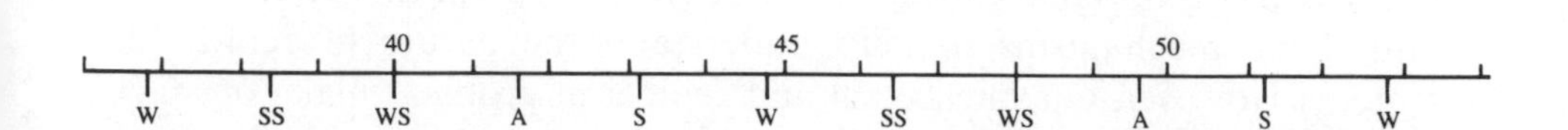

A: AUTUMN S: SPRING W: WINTER SS: SPRING-SUMMER WS: WINTER-SPRING

Figure 82. Small-pulse model for population synchronization.

larch budmoth (*Zeiraphera diniana*) decimation required about 7 years, a peak would occur every 8 years, which spectral analysis (Finerty, unpublished) of Auer's (1977) data for the Swiss Oberengadin from 1949 to 1976 suggests.

Of course, it cannot too often be stated that, at this level, these are only numbers, and until these numbers can be clearly connected to natural phenomena (which could be factors causing sudden increases, or factors precipitating declines; factors affecting herbivores directly, or factors affecting forage recovery), they must be viewed as merely being rather interesting. But they are rather interesting.

Is there any way of trying to test the validity of a short-term cycle affecting long-term oscillations? One might begin by examining plants connected with 4-year-cycle foragers, since the time scale of these cycles is more approachable than that for 10-year periods, and because hormonal studies on animals in natural habitats are singularly difficult, since the hormonal response produced could be quite small compared to what one can accurately measure. These plants could be approached by realizing that the qualitative composition of plant material growing in the early spring often depends on nutrients stored and buds formed in the previous summer (Kalela, 1962) and, over several cycles, one might be able to deduce whether or not a pattern might exist. Yet even here, the results may be equivocal, owing to difficulties of measuring nutrient availability relative to prey density, and one might ultimately be grateful for theories that allow extrapolation of solutions from things that *can* be measured.

Concluding Remarks

Mathematical theory and biological fact suggest that the interrelationships among the various components of boreal and tundra ecosystems provide the setting for natural periods of oscillation which can be seen to be characteristic of the life forms and environments of these systems. These interrelationships can be revealed and explored through mathematical tools that can direct our attention to areas that might reveal underlying mechanisms to help understand the nature of ecological systems. However, our intellectual and ecological probings make one fact very clear: unlike the history of physics, where the data came first and theoretical tools were developed to try to comprehend observed facts, theoretical tools in ecological sciences are far in excess of what we know. This has produced many situations where, as Evelyn Hutchinson (1975) has noted, "experimental observations have been so much harder to obtain than theoretical predictions that the relations between the two have been unhappy."

However one theoretically approaches an ecological question, the most important consequence will be to try to understand the relations of mathematical analyses to biologically interesting things, stressing the need to anchor oneself in concrete biological fact while exploring abstract mathematical schemata. It is my hope that this volume, which has tried to "uncover possibilities by any kind of theoretical analysis that proves helpful," has demonstrated on a small scale the great value of the tools at hand, along with the necessary caveat: the importance of viewing these tools as means and not ends, for in the end one inevitably verifies the observation of Johann Goethe:

> Hypotheses are scaffoldings that one erects in advance of the building and that one takes down when the building is finished. The worker cannot do without them. But he must be careful not to mistake the scaffolding for the building.

Appendix

We consider here a model suggested by Hutchinson and Deevey (1949) to show the pseudoperiodicities that might be expected whenever rapid increase or decrease of a population depends not only on the size of the breeding stock, but also on the occurrence of a critical meteorological condition with a definite probability in the region of interest. The distance between maxima in one region of the series might have an average value near that for another section of the series, but no truly periodic behavior would be expected. The numbers are generated by rolling a pair of dice (simulated on a computer; Finerty, 1972) and subtracting seven, the most probable throw. The resulting number (y) expresses the deviation from the most probable ecological conditions. Taking an initial value for the series of $N_0 = 1.1$, and assuming logarithmic population growth, the effect of this deviation on the progeny of the breeding stock is expressed by successive multiplication by N^y. The resulting curve (fig. 83) looks very much like that observed in natural populations. The autocorrelations, however, do not suggest any periodicity, although the correlation between adjacent observations and the dependence of each observation on all those preceding are apparent. The spectrum is most powerful at low frequencies, indicating the mean shift evident in the original data series, but there is again no suggestion of oscillation.

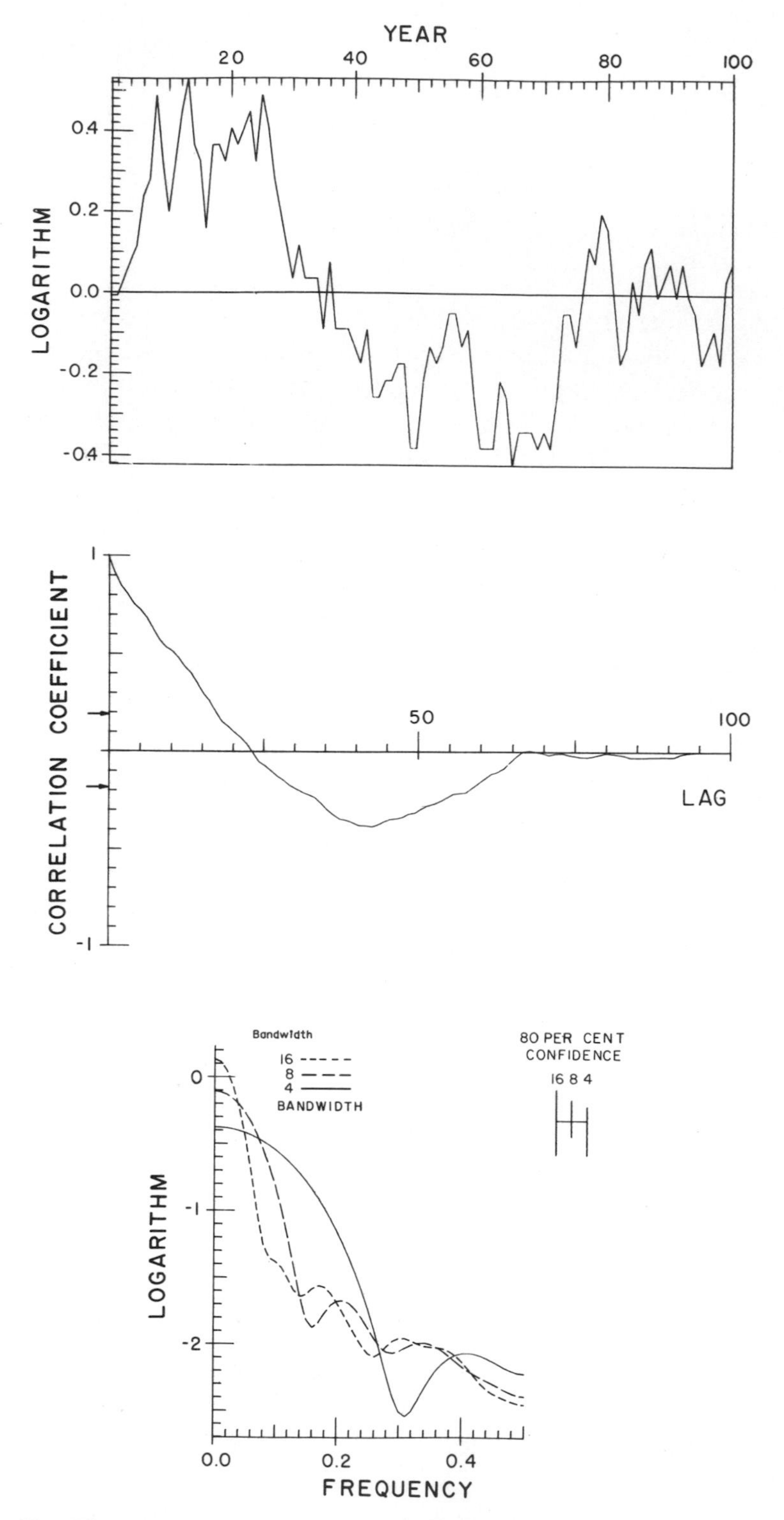

Figure 83. Time-series analysis of a pseudoperiodic model suggested by Hutchinson and Deevey (1949).

References and Bibliography

Alexander, R. D., and T. E. Moore. 1962. The evolutionary relationships of 17-year and 13-year cicadas, and three new species (Homoptera, Cicadidae, Magicicada). Misc. Publ. Mus. Zool. Univ. Mich. 121:5–59.

Andersen, J. 1957. Studies in Danish hare-populations I. Population fluctuations. Dan. Rev. Game Biol. 3:85–131.

Andersen, N. 1974. On the calculation of filter coefficients for maximum entropy spectral analysis. Geophysics 39:69–72.

Anderson, P. 1970. Ecological structure and gene flow in small mammals. Symp. Zool. Soc. Lond. 26:299–325.

Anderson, R. M. 1928. The fluctuation in the population of wild mammals and the relationship of this fluctuation to conservation. Can.-Field Nat. 42:189–91.

———. 1934. Mammals of the eastern Arctic and Hudson's Bay. In: Canada's Eastern Arctic, pp. 67–108. Ottawa.

Anderson, R. M., and A. L. Rand. 1945. The varying lemming (genus *Dicrostonyx*) in Canada. J. Mammal. 26:301–06.

Andrews, R. V. 1968. Daily and seasonal variation in adrenal metabolism of the brown lemming. Physiol. Zool. 41:86–94.

———. 1970. Effects of climate and social pressure on the adrenal glands of lemmings, voles and mice. Acta Endocrinol. 65:639–44.

Andrews, R. V., K. Ryan, R. Strobehn, and M. Rayn-Kline. 1975. Physiological and demographic profiles of brown lemmings during their cycle of abundance. Physiol. Zool. 48:64–83.

Andrews, R. V., and R. Strobehn. 1971. Endocrine adjustments in a wild lemming population during the 1969 summer season. Comp. Biochem. Physiol. 38A:183–201.

Arbib, M. A. 1973. Automata theory in the context of theoretical neurophysiology *in* R. Rosen, ed. Supercellular Systems. Foundations of Mathematical Biology, vol. 3, pp. 228–29. Academic Press, New York.

Audubon, J. J. 1854. The Quadrupeds of North America. V. G. Audubon, New York.

Auer, C. 1975. Jahrliche und langfrustige Dichteveranderungen bei Larchenwickler-Populationen (*Zeiraphera diniana* Gn.) ausserhalb des Optimungebietes. Mitt. Schweiz. Entomol. Ges. 48:47–58.

———. 1977. Dynamik von Larchenwickler-Populationen langs des Alpenbogens. Mitt. Eidg. Anst. Forstl. Versuchwes. 53:75–92.

Aumann, G. D. 1965. Microtus abundance and soil sodium levels. J. Mammal. 46:594–604.

Bailey, R. A. 1946. Reading rabbit population cycles from pines. Wis. Conserv. Bull. 11(7):14–17.

Baltensweiler, W. 1968. The cyclis population dynamics of the grey larch Trotrix, *Zeiraphera griseana* Hubner (= *Semasia diniana* Guenee) (Lepidoptera: Tortricidae). In: T. R. E. Southwood, ed. Insect Abundance, pp. 88–97. Blackwell Scientific Publications, Oxford.

———. 1971. The relevance of changes in the composition of larch bud-moth populations for the dynamics of its numbers. In: den Boer and Gradwell, eds., 1971, pp. 208–19.

———. 1977. Colour-polymorphism and dynamics of larch bud moth populations (*Zeiraphera diniana* Gn., Lep. Tortricidae). Mitt. Schwiez. Entomol. Ges. 50 : 15–23.

Baltensweiler, W., G. Benz, P. Bovey, and V. Delucchi. 1977. Dynamics of larch bud moth populations. Annu. Rev. Entomol. 22:79–100.

Banfield, A. W. F. 1974. The Mammals of Canada. Univ. of Toronto Press, Toronto.

Bangs, O. 1898. A list of the mammals of Labrador. Am. Nat. 32:489–507.

Barnett, S. A. 1964. Social stress. In: J. D. Carthy and C. L. Duddington, eds. Viewpoints in Biology, vol. 3, pp. 170–218. Butterworth, London.

Barry, R. G., and F. K. Hare. 1974. Arctic climate. In: Ives and Barry, 1974, pp. 17–54.

Barry, R. G., and J. D. Ives. 1974. Introduction. In: Ives and Barry, 1974, pp. 1–13.

Bartlett, M. S. 1950. Periodogram analysis and continuous spectra. Biometrica 37:1–16.

———. 1954. Problèmes de 1' analyse spectrale des series temporelles stationnaires. Publ. Inst. Stat. Univ. Paris 3:119–34.

Bartlett, M. S., and R. W. Hiorns, eds. 1973. The Mathematical Theory of the Dynamics of Biological Populations. Academic Press, New York.

Batzli, G. O. 1975. The role of small mammals in arctic ecosystems. In: Golley et al., 1975a, pp. 243–68.

Batzli, G. O., and F. A. Pitelka. 1975. Vole cycles: test of another hypothesis. Am. Nat. 109 : 482–87.

Beddington, J. R., C. A. Free, and J. H. Lawton. 1975. Dynamic complexity in predator–prey models framed in difference equations. Nature 255 : 58–60.

———. 1976. Concepts of stability and resilience in predator–prey models. J. Anim. Ecol. 45:791–816.

Bell, R. 1884. Observations on the geology, mineralogy, zoology, and botany of the Labrador coast. Rep. Geol. Nat. Hist. Surv. Mus. Can. 1882–84. 1–62 DD. (1) 16DD.

———. 1895. The Labrador peninsula. Scott. Geogr. Mag. 11 : 335–61.

Berger, P. J., E. H. Sanders, P. D. Gardner, and N. C. Negus. 1977. Phenolic plant compounds functioning as reproductive inhibitors in *Microtus montanus*. Science 195 : 575–77.

Bethune, W. C. 1935. Canada's Eastern Arctic. Patenaude, Ottawa.

Bhargava, S. C., and R. P. Saxena. 1977. Stable periodic solutions of the reactive-diffusive Volterra system of equations. J. Theor. Biol. 67 : 399–406.

Bider, J. R. 1961. An ecological study of the hare *Lepus americanus*. Can. J. Zool. 39:81–103.

Bigger, M. 1973. An investigation by Fourier analysis into the interaction between coffee leaf-miners and their larval parasites. J. Anim. Ecol. 42:417–34.

Bliss, L. C., G. M. Courtin, D. L. Pattie, R. R. Riewe, D. W. A. Whitfield, and P. Widden. 1973. Arctic tundra ecosystems. Annu. Rev. Ecol. Syst. 4:359–99.

Bloomfield, P. 1976. Fourier Analysis of Time Series: An Introduction. Wiley, New York.

den Boer, P. J., and G. R. Gradwell, eds. 1971. Dynamics of Populations. Proc. Adv. Study Inst. Dynamics Numbers Popul., Oosterbeek, The Netherlands, 1970. Wageningen, Centre for Agricultural Publications and Documents, Wageningen, The Netherlands.

Botkin, D. B., P. A. Jordan, A. S. Dominski, H. S. Lowendorf, and G. E. Hutchinson. 1973. Sodium dynamics in a northern ecosystem (moose, wolves, plants). Proc. Natl. Acad. Sci. U.S.A. 70 : 2745–48.

Bowden, J. 1869. The Naturalist in Norway. Reeve, London.

Box, G. E. P., and G. M. Jenkins. 1970. Times Series Analysis: Forecasting and Control. Holden-Day, San Francisco.

Braestrup, F. W. 1940. The periodic die-off in certain herbivorous mammals and birds. Science 92:354–55.

———. 1941. A study of the arctic fox in Greenland. Medd. Grøn. 131 (4):1–101.

Brain, P. F. 1971. The physiology of population limitation in rodents— a review. Commun. Behav. Biol. 6:115–23.

Brand, C. J., L. B. Keith, and C. A. Fischer. 1976. Lynx responses to changing snowshoe hare densities in central Alberta. J. Wildl. Manage. 40 : 416–28.

Brand, C. J., R. H. Vowles, and L. B. Keith. 1975. Snowshoe hare mortality monitored by telemetry. J. Wildl. Manage. 39 : 741–47.

Brillinger, D. R. 1969. A search for a relationship between monthly sunspot numbers and certain climatic series. Bull. Inst. Int. Stat. 43 : 293–306.

———. 1975. Time Series: Data Analysis and Theory. Holt, Rinehart and Winston, New York.

van den Brink, F. H. 1967. Guide des Mammifères d'Europe. Delachaux et Niestle, Suisse.

Bryson, R. A., and F. K. Hare. 1974. Climates of North America. World Survey of Climatology, Vol. 2. Elsevier, Amsterdam.

Buckley, J. L. 1954. Animal population fluctuations in Alaska—a history. Trans. N. Am. Wildl. Conf. 19 : 338–57.

Bulmer, M. G. 1974. A statistical analysis of the 10-year cycle in Canada. J. Anim. Ecol. 43:701–18.

———. 1975a. Phase relations in the ten-year cycle. J. Anim. Ecol. 44:609–21.

———. 1975b. The statistical analysis of density dependence. Biometrics 31:901–11.

———. 1976. The theory of prey–predator oscillations. Theor. Popul. Biol. 9:137–50.

Burton, M. 1962. Systematic Dictionary of Mammals of the World. Museum Press, London.

Buss, I. O. 1950. (In discussion.) Trans. N. Am. Wildl. Conf. 15:382.

Butler, L. 1942. Fur cycles and conservation. Trans. N. Am. Wildl. Conf. 7:463–72.

———. 1945. Distribution and genetics of the color phases of the red fox in Canada. Genetics 30:39–50.

———. 1947. The genetics of the color phases of the red fox in the Mackenzie River locality. Can. J. Res., D Zool. Sci. 25:190–215.

———. 1950. Canada's wild fur crop. Beaver, Dec. 1950.

———. 1951. Population cycles and color phase genetics of the colored fox in Quebec. Can. J. Zool. 29:24–41.

———. 1953. The nature of cycles in populations of Canadian mammals. Can. J. Zool. 31:242–62.

———. 1962. Periodicities in the annual muskrat population figures for the province of Saskatchewan. Can. J. Zool. 40:1277–87.

Butler, W. F. 1872. The Great Lone Land. Sampson, Low, Marston, Low and Searle, London.

———. 1873. The Wild Northland. Sampson, Low, Marston, Low and Searle, London.

Cabot, W. B. 1912. In Northern Labrador, Appendix: Mice, pp. 287–92. Badger, Boston.

Cary, J. R., and L. B. Keith. 1979. Reproductive changes in the 10-year cycle of snowshoe hares. Can. J. Zool. 57:375–90.

Caswell, H. 1972. A simulation study of a time lag population model. J. Theor. Biol. 34:419–39.

Charnov, E. L., and J. P. Finerty. 1978. Vole population cycles: a case for kin selection? Submitted for publication.

Chatfield, C. 1975. The Analysis of Time Series: Theory and Practice. Chapman & Hall, London.

Chesemore, D. L. 1968. Notes on the food habits of Arctic foxes in northern Alaska. Can. J. Zool. 46:1127–30.

Chitty, D. 1952. Mortality among voles (*Microtus agrestis*) at Lake Vyrnwy, Montgomeryshire in 1936–39. Phil. Trans. R. Soc. Lond. B Biol. Sci. 236:505–52.

———. 1957. Self-regulation of numbers through changes in viability. Cold Spring Harbor Symp. Quant. Biol. 22:277–80.

———. 1959. A note on shock disease. Ecology 40:728–31.

———. 1960. Population processes in the vole and their relevance to general theory. Can. J. Zool. 38:99–113.

———. 1967. The natural selection of self-regulatory behavior in animal populations. Proc. Ecol. Soc. Aust. 2:51–78.

———. 1969. Regulatory effects of a random variable. Am. Zool. 9:400.

———. 1970. Variation and population density. Symp. Zool. Soc. Lond. 26:327–33.

Chitty, H. 1950. The snowshoe rabbit enquiry, 1946–1948. J. Anim. Ecol. 19:15–20.

———. 1961. Variations in the weight of the adrenal glands of the field vole, *Microtus agrestis*. J. Endocrinol. 22:387–93.

Christian, J. J. 1950. The adreno-pituitary system and population cycles in mammals. J. Mammal. 31:247–59.

———. 1961. Phenomena associated with population density. Proc. Natl. Acad. Sci. U.S.A. 47:428–49.

———. 1970. Social subordination, population density, and mammalian evolution. Science 168:84–90.

———. 1971a. Population density and reproduction efficiency. Biol. Reprod. 4:248–94.

———. 1971b. Fighting, maturity, and population density in *Microtus pennsylvanicus* J. Mammal. 52:556–67.

———. 1975. Hormonal control of population growth. In: B. E. Eleftheriou and R. L. Sprott, eds. Hormonal Correlates of Behavior, vol. 1, pp. 205–74. Plenum Press, New York.

Christian, J. J., and D. E. Davis. 1964. Endocrines, behavior and population. Science 146:1550–60.

Christian, J. J., J. A. Lloyd, and D. E. Davis. 1965. The role of endocrines in the self-regulation of mammalian populations. Recent Progr. Horm. Res. 21:501–78.

Clarke, C. H. D. 1936. Fluctuations in numbers of ruffed grouse. *Bonasa umbellus* (linne), with special reference to Ontario. Univ. Toronto Stud. Biol. Ser. 41.

Clopper, C. J., and E. S. Pearson. 1934. The use of confidential or fiducial limits illustrated in the case of the binomial. Biometrika 26:404–13.

Clough, G. C. 1965. Lemmings and population problems. Am. Sci. 53:199–212.

———. 1968. Social behavior and ecology of Norwegian lemmings during a population peak and crash. Pap. Norw. State Game Res. Inst. Ser. 2, 28:1–49.

Clulow, F. V., and J. R. Clarke. 1968. Pregnancy block in *Microtus agrestis* an induced ovulator. Nature 219:511.

Cody, M. L., and J. M. Diamond, eds. 1975. Ecology and Evolution of Communities. Belknap Press of Harvard University Press, Cambridge, Mass.

Cohen, J. E. 1970. A Markov contingency-table model for replicated Lotka-Volterra systems near equilibrium. Am. Nat. 104:547–60.

Cole, L.C. 1951. Population cycles and random oscillations. J. Wildl. Manage. 15:233–52.

———. 1954. Some features of random population cycles. J. Wildl. Manage. 19(1):1–24.

Collett, R. 1878. On *Myodes lemmus* in Norway. J. Linn. Soc. Lond. Zool. 13 : 329–34.

———. 1895. *Myodes lemmus*: its habits and migration in Norway. Forh. VidenskSelsk. Krist. 3:1–62.

———. 1911–12. Norges Pattedyr. Oslo, Norway.

Collin, R. 1954. The Theory of Celestial Influence. Weiser, New York.

Cottam, C. 1936. Food of arctic birds and mammals collected by the Bartlett expeditions of 1931, 1932, and 1933. J. Wash. Acad. Sci. 26: 165–77.

Cottam, C., and H. C. Hanson. 1938. Food habits of some Arctic birds and mammals. Field Mus. Nat. Hist. Publ. Zool. Ser. 20(31):405–26.

Coues, E. 1877. Fur-bearing mammals: A Monograph of North American Mustelidae. U.S. Dept. Inter. Misc. Publ. 8.

———. 1897. The Manuscript Journals of Alexander Henry and David Thompson, 1799–1814. London.

Cowan, I. M. 1938. The fur trade and fur cycle: 1825–1857. B. C. Hist. Quart. 1938: 19–30.

Cox, W. T. 1936. Snowshoe rabbit migration, tick infestation, and weather cycles. J. Mammal. 17:216–21.

Criddle, N. 1931. Some natural factors governing the fluctuations of grouse in Manitoba. Can. Field Nat. 44:77–80.

———. 1932. The correlation of sunspot periodicity with grasshopper fluctuation in Manitoba. Can. Field Nat. 46:195–99.

Criddle, S. 1938. A study of the snowshoe rabbit. Can. Field Nat. 52:31–40.

Cross, E. C. 1940. Periodic fluctuations in numbers of the red fox in Ontario. J. Mammal. 21:294–306.

Cunningham, W. J. 1954. A non-linear differential-difference equation of growth. Proc. Natl. Acad. Sci. U.S.A. 40:708–13.

Curry-Lindahl, K. 1962. The irruption of the Norway lemming in Sweden during 1960. J. Mammal. 43(2):171–84.

D'Ancona, U. 1954. The Struggle for Existence. Brill, Leiden.

Davis, W. B. 1944. Geographic variation in brown lemmings (genus *Lemmus*). Murrelet 25 : 19–25.

Dennis, J., and P. L. Johnson, 1970. Shoot and rhizome-root standing crops of tundra vegetation at Barrow, Alaska. Arct. Alp. Res. 2:253–66.

Diereville, N. de. 1933. Relation of the voyage to Port Royal in Acadia or New France. Mrs. C. Webster, trans., J. C. Webster, ed. Toronto, The Champlain Society (first edition 1708, Rouen).

Dolbeer, R. A., and W. R. Clark. 1975. Population ecology of snowshoe hares in the central Rocky Mountains. J. Wildl. Manage. 39 : 535–49.

Dominion Bureau of Statistics. June, 1970. Fur Production. Cat. 23–207.

Ellerman. J. R., and T. C. S. Morrison-Scott. 1951. Checklist of Palearctic and Indian Mammals, 1758–1946. British Museum, London.

Elton, C. S. 1924. Periodic fluctuations in the numbers of animals: their causes and effects. J. Exp. Biol. 2:119–63.

———. 1925. Plague and the regulation of numbers in wild mammals. J. Hyg. 24:138–63.
———. 1930. Animal Ecology and Evolution. Clarendon Press, Oxford.
———. 1931a. Epidemic sledge dogs and their relation to arctic fox. Can. J. Res. 5:673–92.
———. 1931b. The study of epidemic diseases among wild animals. J. Hyg. 31:435–56.
———. 1931c. Matamek Conference on Biological Cycles. Matamek Factory, Canadian Labrador.
———. 1933. The Canadian snowshoe rabbit enquiry, 1931–1932. Can. Field. Nat. 47:63–69, 84–86.
———. 1942. Voles, Mice and Lemmings. Clarendon Press, Oxford.
———. 1958. The Ecology of Invasions by Animals and Plants. Wiley, New York.
Elton, C. S., E. B. Ford, J. R. Baker, and A. D. Gardner. 1931. The health and parasites of a wild mouse population. Proc. Zool. Soc. Lond. 1931:657–721.
Elton, C. S., and M. Nicholson. 1942a. Fluctuations in numbers of the muskrat (*Ondatra zibethica*) in Canada. J. Anim. Ecol. 11:96–126.
———. 1942b. The ten-year cycle in numbers of the lynx in Canada. J. Anim. Ecol. 11:215–44.
Errington, P. L. 1954. On the hazards of overemphasizing numerical fluctuations in studies of "cyclic" phenomena in muskrat populations. J. Wildl. Manage. 18:66–90.
———. 1963. Muskrat Populations. Iowa State Univ. Press, Ames, Iowa.
Evans, F. C. 1942. Study of a small mammal population in Bagley Wood, Berkshire. J. Anim. Ecol. 11:182–97.
Finerty, J. P. 1972. Cyclic fluctuations in biological systems: a revaluation. Ph.D. dissertation, Yale University, New Haven.
———. 1978. Cycles in lynx. In press: Am. Nat. 114:453–55.
Fisher, R. A. 1929. Test of significance in harmonic analysis. Proc. R. Soc. Lond. A125:54–59.
Fitzgerald, B. M. 1977. Weasel predation on a cyclic population of the montane vole (*Microtus montanus*) in California. J. Anim. Ecol. 46:367–97.
Fox, J. F. 1978. Forest fires and the snowshoe hare–Canada lynx cycle. Oecologia 31:349–74.
Frank, F. 1957. The causality of microtine cycles in Germany. J. Wildl. Manage. 21:113–21.
Freeland, W. J. 1974. Vole cycles: another hypothesis. Am. Nat. 108:238–45.
Fuller, W. A. 1967. Écologie hivernale des lemmings et fluctuations de leurs populations. Terre Vie 2:97–115.
———. 1969. Changes in numbers of three species of small rodent near Great Slave Lake, N. W. T. Canada, 1964–1967, and their significance for general population theory. Zool. Fenn. Ann. 6:113–44.

Fuller, W. A., P. G. Kevan, eds. 1970. Productivity and Conservation in Northern Circumpolar Lands. Gresham Press, Surrey, England.

Fuller, W. A., L. L. Stebbins, and G. R. Dyke. 1969. Over-wintering of small mammals near Great Slave Lake, Northern Canada. Arctic 22 : 34–55.

Gaines, M. S., and C. J. Krebs. 1971. Genetic changes in fluctuating vole populations. Evolution 25 : 702–23.

Garfinkel, D. 1967. A simulation study of the effect on simple ecological systems of making rate of increase of population density-dependent. J. Theor. Biol. 14:46–58.

Garten, C. T., Jr. 1976. Relationships between aggressive behavior and genic heterozygosity in the oldfield mouse, *Peromyscus polionotus*. Evolution 30 : 59–72.

Gause, G. F. 1934. The Struggle for Existence. The Williams and Wilkins Co., Baltimore.

Gause, G. E., N. P. Smaragodova, and A. A. Witt. 1936. Further studies on interaction between predators and prey. J. Anim. Ecol. 5:1–18.

Gerety, E. J., J. M. Wallace, and C. S. Zerefos. 1977. Sunspots, geomagnetic indices and the weather: a cross-spectral analysis between sunspots, geomagnetic activity and global weather data. J. Atmos. Sci. 34:673–77.

Gilpin, M. E. 1973. Do hares eat lynx? Am. Nat. 107:727–30.

Glass, L. 1975. Classification of biological networks by their qualitative dynamics. J. Theor. Biol. 54:85–107.

Gmelin, J. G. 1760. Animalium quorundam quadrupedum descriptio. Novi Comment. Acad. Sci. Imp. Petrop. 5:338–72.

Godfrey, M. D. 1974. Computational methods for time series. Bull., Institute for Mathematics and Its Applications. 10 : 224–27.

Goel, N. S., S. C. Maitra, and E. W. Montroll. 1971. On the Volterra and other non-linear models of interacting populations. Rev. Mod. Phys. 43:231–76.

Golley, F. B., K. Petrusewicz, and L. Ryszkowski, eds. 1975a. Small Mammals: Their Productivity and Population Dynamics. International Biological Programme, vol. 5. Cambridge Univ. Press, Cambridge, England.

Golley, F. B., L. Ryszkowski, and J. T. Sokur. 1975b. The role of small mammals in temperate forests, grasslands and cultivated fields. In: Golley et al: 1975a, pp. 223–41.

Goodman, D. 1975. The theory of diversity-stability relationships in ecology. Quart. Rev. Biol. 50:237–66.

Goodman, L. A. 1967. On the reconciliation of mathematical theories of population growth. J. R. Stat. Soc., Ser. A 130:541–53.

Grange, W. B. 1949. The Way to Game Abundance. Scribner's, New York.

———. 1965. Fire and tree growth relationships to showshoe rabbits. Proc. Annu. Tall Timbers Fire Ecol. Conf. 4:110–25.

Grant, P. R. 1978. Dispersal in relation to carrying capacity. Proc. Natl. Acad. Sci. U.S.A. 75:2854–58.

Green, R. G., and C. A. Evans. 1940a. Studies on a population cycle of snowshoe hares on the Lake Alexander area. I. Gross annual censuses, 1932–1939. J. Wildl. Manage. 4 : 220–38.

———. 1940b. Studies on a population cycle of snowshoe hares on the Lake Alexander area. II. Mortality according to age groups and seasons. J. Wildl. Manage. 4:267–78.

———. 1940c. Studies on a population cycle of snowshoe hares on the Lake Alexander area. III. Effects of reproduction and mortality of young hares on cycle. J. Wildl. Manage. 4:347–58.

Green, R. G., and C. L. Larson. 1938a. A description of shock disease in the snowshoe hare. Am. J. Hyg. 28(2): 190–212.

———. 1938b. Shock disease and the snowshoe hare cycle. Science 87:298–99.

Green, R. G., C. L. Larson, and J. F. Bell. 1939. Shock disease as the cause of the periodic decimation of the snowshoe hare. Am. J. Hyg. 30B(3):83–102.

Gross, A. O. 1947. Cyclic invasions of the snowy owl. Auk 64 : 584–601.

Guckenheimer, J., G. F. Oster, and A. Ipaktchi. 1977. The dynamics of density-dependent population models. J. Math. Biol. 4 : 101—47.

Gudmundsson, F. 1960. Some reflections on Ptarmigan cycles in Iceland. Proc. Int. Ornithol. Congr. 12:259–65.

Gurney, W. S. C., and R. M. Nisbet. 1978. Predator–prey fluctuations in patchy environments. J. Anim. Ecol. 47:85–102.

Hall, E. R., and K. R. Kelson, 1959. The Mammals of North America, vol. 2. Ronald Press, New York.

Hamann, J. R., and L. M. Bianchi. 1970. Stochastic population mechanics in the relational systems formalism: Lotka–Volterra ecological dynamics. J. Theor. Biol. 28:175–84.

Hamilton, W. D. 1964. The genetical evolution of social behavior. J. Theor. Bio. 12:1–52.

———. 1971. Selection of selfish and altruistic behavior in some extreme models. In: J. F. Eisenberg and W. S. Dillon, eds. Man and Beast: Comparative Social Behavior, pp. 57–91. Smithsonian Institution Press, Washington, D.C.

———. 1972. Altruism and related phenomena. Ann. Rev. Ecol. Syst. 3:193–232.

Hannan, E. J. 1960. Time Series Analysis. Wiley, New York.

Hare, F. K., and J. E. Hay. 1974. The climate of Canada and Alaska. In: Bryson and Hare, 1974, pp. 49–192.

Hare, F. K., and J. C. Ritchie. 1972. The boreal bioclimates. Geogr. Rev. 62:333–65.

Harper, J. L. 1977. Population Biology of Plants. Academic Press, New York.

Hassell, M. P., J. H. Lawton, and R. M. May. 1976. Patterns of dynamical behavior in single-species populations. J. Anim. Ecol. 45:471–86.

Hawkes, E. W. 1916. The Labrador Eskimo. Can. Dept. Mines Geol. Surv. Mem. 91 (Anthropol. Ser. 14):1–235.

Heller, R. 1978. Two predator–prey difference equations considering delayed population growth and starvation. J. Theor. Biol. 70:401–13.

Henshaw, J. 1966. Mass movements by snowshoe rabbits (*Lepus americanus*). Can. Field. Nat. 80:181.

Hewitt, C. G. 1921. The Conservation of the Wild Life of Canada. Scribner's, New York.

Hilborn, R., and C. J. Krebs. 1976. Fates of disappearing individuals in fluctuating populations of *Microtus townsendii.* Can. J. Zool. 54:1507–18.

Hind, H. Y. 1860. Narrative of the Canadian Red River Expedition of 1857 and of the Assinniboine and Saskatchewan exploring expedition of 1858, 2 vols. Longman, Green, Longman and Roberts, London.

———. 1863. Explorations in the Interior of the Labrador Peninsula, the country of the Montaquais and Nasquapec Indians. Longman, Green, Longman, Roberts and Green, London.

Hoffmann, R. S. 1958. The role of predators in "cyclic" declines of grouse populations. J. Wildl. Manage. 22:317–19.

———. 1974. Terrestrial vertebrates. In: Ives and Barry, 1974, pp. 475–568.

Holling, C. S. 1959. The components of predation as revealed by a study on small-mammal predation of the European pine sawfly. Can. Entomol. 91:293–320.

———. 1965. The functional response of predators to prey density and its role in mimicry and population regulation. Mem. Entomol. Soc. Can. 45:3–60.

———. 1973. Resilience and stability of ecological systems. Annu. Rev. Ecol. Syst. 4:1–23.

Hoppensteadt, F. C. 1976. Mathematical Methods of Population Biology. Courant Institute of Mathematical Sciences, New York Univ., New York.

Howell, A. B. 1923. Periodic fluctuations in the numbers of small mammals. J. Mammal. 4:149–55.

Huffaker, C. B. 1958. Experimental studies on predation: dispersion factors and predator–prey oscillations. Hilgardia 27:343–83.

Hustich, I. 1968. Plant geographical regions. Sømme 1968:62–70.

Hutchinson, G. E. 1942. Nati sunt mures, et facta est confusio. Quart. Rev. Biol. 17:354–57.

———. 1948. Circular causal systems in ecology. Ann. N.Y. Acad. Sci. 50:221–46.

———. 1949. Marginalia. Am. Sci. 37:592–618.

———. 1954. Theoretical notes on oscillatory populations. J. Wildl. Manage. 18:107–09.

———. 1957. Concluding remarks. Cold Spring Harbor Symp. Quant. Biol. 22:415–27.

———. 1959. Homage to Santa Rosalia, or why are there so many kinds of animals? Am. Nat. 93:145–59.

———. 1975. Variations on a theme by Robert MacArthur. In: M. L. Cody and J. M. Diamond, eds., 1975, pp. 492–521.

———. 1978. An Introduction to Population Ecology. Yale Univ. Press, New Haven.

Hutchinson, G. E., and E. S. Deevey, Jr. 1949. Ecological Studies on Populations. Survey of Biological Progress, vol. 1. Academic Press, New York.

Hutton, S. K. 1912. Among the Eskimos of Labrador. London.

Ives, J. D. 1974. Biological refugia and the nunatek hypothesis. In: Ives and Barry, 1974, pp. 605–36.

Ives, J. D., and R. G. Barry. 1974. Arctic and Alpine Environments. Methuen, London.

Janzen, D. H. 1976. Why bamboos wait so long to flower. Annu. Rev. Ecol. Syst. 7:347–91.

Jenkins, G. M., and D. G. Watts. 1968. Spectral Analysis and Its Applications. Holden-Day, San Francisco.

Jenness, D. 1932. The Indians of Canada. Bull. Natl. Mus. Can. 65:273.

Johannessen, T. W. 1970. The climate of Scandinavia. In: Wallen, ed., 1970, pp. 23–79.

Johnsen, S. 1929. Rovdyr-og rovfugl-statistikken i Norge. Mergens Mus. Arbok, Naturvidensk. Rekke 2.

Jones, J. W. 1914. Fur-farming in Canada (2d edition, revised and enlarged). Canada, Commission of Conservation, Committee on Fisheries, Game and Fur-bearing Animals. Ottawa.

Jones, R. H. 1965. A reappraisal of the periodogram in spectral analysis. Technometrics 7:531–42.

———. 1971. Spectrum estimation with missing observations. Ann. Inst. Stat. Math. 23:387–98.

Kalela, O. 1949. Zur Periodizität im Massenwechsel arktisch-Lochborealer Kleinnager. Arch. Soc. Zool. Bot. Fenn. Vanamo 3:169–78.

———. 1951. Einige Konsequenzen aus der regionalen Intensitätsvariation in Massenwechsel der Saugetieren und Vogel. Ann. Zool. Soc. Zool. Bot. Fenn. Vanamo 14.

———. 1954. Population okologishe Gesichtspunkte zur Entstehung des Vogelzuges. Ann. Zool. Soc. Bot. Fenn. Vanamo 16.

———. 1957. Regulation of reproduction rate in subarctic populations of the vole *Clethrionomys reufocanus* (Sund.). Suom. Tied. Toimit. Ser. R, Biol. 34.

———. 1962. On the fluctuations in the numbers of arctic and boreal small rodents as a problem of population biology. Ann. Acad. Sci. Fenn. Ser. A, IV. Biol. 66:1–38.

Kalela, O., and T. Koponen. 1971. Food consumption and movements of the Norwegian lemmings in areas characterized by isolated fells. Ann. Zool. Fenn. 8:80–84.

Kalela, O., T. Koponen, E. Lind, U. Skaren, and J. Tast. 1961. Seasonal change of habitat in the Norwegian lemming, *Lemmus lemmus* (L.). Ann. Acad. Sci. Fenn. Ser. A, IV. Biol. 55:1–72.

Kalm, Pehr. 1756. Beskfning pa et slags Grashopper uti Norra America. Kongl. Sven. Vetensk. Acad. Handl. 17:101–16. In: J. J. Davis. 1953. Pehr Kalm's description of the periodical cicada, Magicicada septendecim L. Ohio J. Sci. 53:138–42.

Keith, L. B. 1963. Wildlife's Ten-Year Cycle. Univ. of Wisconsin Press, Madison.

———. 1974. Some features of population dynamics in mammals. Proc. Int. Congr. Game Biol. Stockholm 11:17–58.

Keith, L. B., A. W. Todd, C. J. Brand, R. S. Adamcik, and D. H. Rusch. 1977. An analysis of predation during a cyclic fluctuation of snowshoe hares. Trans. XIIIth Congr. Game Biol. Atlanta, pp. 151–75.

Keith, L. B., and L. A. Windberg. 1978. A demographic analysis of the snowshoe hare cycle. Wildl. Monogr. 58.

Keller, B. L., and C. J. Krebs. 1970. Microtus population biology III. Reproductive changes in fluctuating populations of *M. ochrogaster* and *M. pennsylvanicus* in southern Indiana, 1965–1967. Ecol. Monogr. 40:263–94.

Kennedy, A. W. M. C. 1878. The Arctic Regions and Back in Six Weeks. London.

Kerner, E. H. 1957. A statistical mechanics of interacting biological species. Bull. Math. Biophys. 19:121–46.

———. 1959. Further considerations on the statistical mechanics of biological associations. Bull. Math. Biophys. 21:217–55.

———. 1961. On the Volterra–Lotka principle. Bull. Math. Biophys. 23:141–57.

———. 1964. Dynamical aspects of kinetics. Bull. Math. Biophys. 26:333–49.

———. 1972. Gibbs Ensemble: Biological Ensemble. Gordon and Breach, New York.

Kiepenheuer, K. O., I. Brauer, and C. Harte. 1949. Über die Wirkung von Meterwellen auf das Teilungswachstum der Pflanzen. Naturwissenschaften 36(1):27–28.

Kiester, A. R., and R. Barakat. 1974. Exact solutions to certain stochastic differential equation models of population growth. Theor. Popul. Biol. 6:199–216.

Kilmer, W. L. 1972. On some realistic constraints in prey–predator mathematics. J. Theor. Biol. 36:9–22.

Kimmins, J. P. 1970. Cyclic fluctuations in herbivore populations in northern ecosystems: a general hypothesis. Ph.D. dissertation, Yale Univ., New Haven.

Kohn, P., and R. Tamarin. 1978. Selection at electrophoretic loci for reproductive parameters in island and mainland voles. Evolution 32:15–28.

Kolmogoroff, A. 1936. Sulla teoria di Volterra della lotta per l'Esistenza. Giorn. Inst. Ital. Attuari 7:74–80.

Koopman, L. H. 1974. The Spectral Analysis of Time Series. Academic Press, New York.

Koponen, T. 1970. Age structure in sedentary and migratory populations of the Norwegian lemming, *Lemmus lemmus* (L.), at Kilpisjarvi in 1960. Ann. Zool. Fenn. 7:141–87.

Koponen, T., A. Kokkonen, and O. Kalela. 1961. On a case of spring migration in the Norwegian lemming. Ann. Acad. Sci. Fenn. V. Biol. 52:1–30.

Koshkina, T. V. 1966. On the periodical changes in the numbers of voles (as exemplified by the Kola Peninsula). (In Russian: trans. W. A. Fuller.) Bull. Mosc. Soc. Natl. Biol. Sect. 71:14–26.

———. 1967a. Ecologic differentiation of species: the vole as an example. Acta Theriol. 12:135–63.

———. 1967b. Population regulation of rodents (example: the northern red-backed vole and the Norwegian lemming). Byull. Mosk. O-va. Ispyt. Prir. Otd. Biol. 72:5–20.

———. 1969. Predicting the number of forest voles by ecological indices of their population state. Byull. Mosk. O-va. Ispyt. Prir. Otd. Biol. 74:5–16.

Koskimies, J. 1955. Ultimate causes of cyclic fluctuations in animal populations. Pap. Game Res. 15:1–29.

Krebs, C. J. 1964a. The lemming cycle at Baker Lake, Northwest Territories, during 1959–1962. Arct. Inst. N. Am. Tech. Pap. 15.

———. 1964b. Cyclic variation in skull–body regressions in lemmings. Can. J. Zool. 42:631–43.

———. 1970. Microtus population biology: behavioral changes associated with the population cycles in *M. ochrogaster* and *M. pennsylvanicus.* Ecology 51 : 34–52.

———. 1971. Genetic and behavioral studies of fluctuating vole populations. In: den Boer and Gradwell, eds., 1971, pp. 243–56.

———. 1978. A review of the Chitty hypothesis of population regulation. Can. J. Zool. 56:2463–80.

Krebs, C. J., M. S. Gaines, B. L. Keller, J. H. Myers, and R. H. Tamarin. 1973. Population cycles in small rodents. Science 179 : 34–41.

Krebs, C. J., B. L. Keller, and R. H. Tamarin. 1969. *Microtus* population biology: demographic changes in fluctuating populations of *M. ochrogaster* and *M. pennsylvanicus* in southern Indiana. Ecology 50 : 587–607.

Krebs, C. J., and J. H. Myers. 1974. Population cycles in small mammals. Adv. Ecol. Res. 8 : 267–399.

Krebs, C. J., I. Wingate, J. LeDuc, J. A. Redfield, M. Taitt, and R. Hilborn. 1976. *Microtus* population biology: dispersal in fluctuating populations of *M. townsendii.* Can. J. Zool. 54 : 79–95.

Lack, D. L. 1954a. Cyclic mortality. J. Wildl. Manage. 18:25–37.

———. 1954b. The Natural Regulation of Animal Numbers. Oxford Univ. Press, Oxford.

———. 1966. Population Studies of Birds. Oxford Univ. Press, Oxford.

Lanchester, F. W. 1916. Aircraft in Warfare: The Dawn of the Fourth Arm. Constable, London.

Lane, P., and R. Levins. 1977. The dynamics of aquatic systems 2. The effects of nutrient enrichment of model plankton communities. Limnol. Oceanogr. 22:454–71.

Larsen, J. A. 1974. Treeline: ecology of the northern continental forest border. In: Ives and Barry, 1974, pp. 341–69.

Lauckhart, J. B. 1957. Animal cycles and food. J. Wildl. Manage. 21:230–34.

———. 1962. Wildlife population fundamentals. 27th N. Am. Wildl. Conf. 27:233–42.

Leigh, E. G., Jr. 1965. On the relation between the productivity, biomass, diversity, and stability of a community. Proc. Natl. Acad. Sci. U.S.A. 53(4):777–83.

———. 1968. The ecological role of Volterra's equations. In: M. Gerstenhaber, ed. Some Mathematical Problems. In: Biology, pp. 1–61. American Mathematical Society, Providence, R.I.

———. 1975. Population fluctuations, community stability, and environmental variability. In: Cody and Diamond, 1975, pp. 51–80.

Leopold, A. 1931. Report on a game survey of the north central states. Sporting Arms and Ammunition Manufacturing Institute, Madison, Wis.

———. 1933. Game Management. Scribner's, New York.

Leopold, A., and J. N. Ball. 1931. British and American grouse cycles. Can. Field Nat. 45:162–67.

Leslie, P. H. 1959. The properties of a certain lag type of population growth and the influence of an external random factor on a number of such populations. Physiol. Zool. 32:151–59.

Levikson, B. 1976. Regulated growth in random environments. J. Math. Biol. 3:19–26.

Levin, S. A. 1970. Community equilibria and stability, and an extension of the competitive exclusion principle. Am. Nat. 104:413–32.

Levins, R. 1974. Qualitative analysis of partially specified systems. Ann. N.Y. Acad. Sci. 231:123–38.

———. 1975. Evolution in communities near equilibrium. In: Cody and Diamond, 1975, pp. 16–50.

———. 1979. Coexistence in a variable environment. Am. Nat. 114: December.

Li, T. Y., and J. A. Yorke, 1975. Period three implies chaos. Am. Math. Mon. 82:985–92.

Libby, L. M., and L. J. Pandolfi. 1977. Climate periods in trees, ice and tides. Nature 266:415–17.

Lidicker, W. Z., Jr. 1962. Emigration as a possible mechanism permitting the regulation of population density below carrying capacity. Am. Nat. 96:29–33.

———. 1973. Regulation on numbers in an island population of the California vole, a problem in community dynamics. Ecol. Monogr. 43:271–302.

———. 1975. The role of dispersal in the demography of small mammals. In: Golley et al., 1975a, pp. 103–33.

Lin, J., and P. B. Kahn. 1976. Averaging methods in predator–prey systems and related biological models. J. Theor. Biol. 57:73–102.

Linnaeus, C. 1758. Systema naturae . . . (10th edition). L. Salvii, Holmiae.

Lloyd, L. 1854. Scandinavian Adventures, vol. 2, p. 77.

Lotka, A. J. 1910. Contribution to the theory of periodic reactions. J. Phys. Chem. 14:271–74.

———. 1920. Analytical note on certain rhythmic relations in organic systems. Proc. Natl. Acad. Sci., U.S.A. 6:410–15.

———. 1923. Contribution to the analysis of malaria epidemiology. Am. J. Hyg. 3 (Jan. suppl.):113–21.

———. 1925. Elements of Physical Biology. Williams & Wilkins, Baltimore.
MacArthur, R. H. 1968. Selection for life tables in periodic environments. Am. Nat. 102:381–83.
———. 1969. Species packing, and what interspecies competition minimizes. Proc. Natl. Acad. Sci. U.S.A. 64 : 1369–71.
———. 1970. Species packing and competitive equilibrium for many species. Theor. Popul. Biol. 1:1–11.
MacArthur, R. H., and R. Levins. 1967. The limiting similarity, convergence and divergence of coexisting species. Am. Nat. 101 : 377–85.
MacArthur, R. H., and E. O. Wilson. 1967. The Theory of Island Biogeography. Monogr. Popul. Biol. 1. Princeton Univ. Press, Princeton, N.J.
MacFarlane, R. 1905. Mammals of the Northwest Territory. Smithson. Inst. Proc. U.S. Natl. Mus. 28(1405):673–764.
McLean, J. 1833–36. Published in 1932. Notes of a Twenty-five Years' Service in the Hudson's Bay Territory. W. S. Wallace, ed. The Champlain Society, Toronto.
MacLean, S. F., Jr., B. M. Fitzgerald, and F. A. Pitelka. 1974. Population cycles in Arctic lemmings: winter reproduction and predation by weasels. Arct. Alp. Res. 6 : 1–12.
MacLulich, D. A. 1937. Fluctuations in the numbers of the varying hare. Univ. Toronto Stud. Biol. Ser. 43.
———. 1957. The place of chance in population processes. J. Wildl. Manage. 21(3):293–99.
Macoun, J. 1883. Manitoba and the Great North-West. T. C. Jack, London.
———. and J. M. Macoun. 1909. Catalogue of Canadian birds. Can. Dept. Mines Geol. Surv. Branch Publ. 973.
MacPherson, A. H. 1969. The dynamics of Canadian arctic fox populations. Can. Wildl. Serv. Rep. Ser. 8.
Magnus, Olaus. 1555a. Historia de Gentibus Septentrioralibus, pp. 617–18. Rome.
———. 1555b. trans. 1658. A Compendious History of the Goths, Swedes, and Vandals, and Other Northern Nations. J. Streater, London.
Mair, C., and R. MacFarlane. 1908. Through the Mackenzie Basin. William Briggs, Toronto.
Manniche, A. L. V. 1910. The terrestrial mammals and birds of Northeast Greenland. Medd. Grøn. 45(1) : 1–99.
Marsden, W. 1964. The Lemming Year. Chatto & Windus, London.
May, R. M. 1972. Limit cycles in predator–prey communities. Science 177:900–02.
———. 1973a. Stability and Complexity in Model Ecosystems. Princeton Univ. Press, Princeton, N.J.
———. 1973b. Qualitative stability in model ecosystems. Ecology 54:638–41.
———. 1974a. How many species: some mathematical aspects of the dynamics of populations. In: Lectures of Mathematics in the Life Sciences, vol. 6. The American Mathematical Society, Providence, R.I.

———. 1974b. Biological populations with non-overlapping generations: stable points, stable cycles and chaos. Science 186:645–47.

———. 1975. Biological populations obeying difference equations: stable points, stable cycles and chaos. J. Theor. Biol. 51:511–24.

———. 1976a. Simple mathematical models with very complicated dynamics. Nature 261:459–67.

———. 1976b. Models for single populations. In: R. M. May, ed. Theoretical Ecology: Principles and Applications, pp. 4–25. Saunders, Philadelphia.

———. 1976c. Models for two interacting populations. In: R. M. May, ed. Theoretical Ecology: Principles and Applications, pp. 9–70. Saunders, Philadelphia.

———. 1977. Thresholds and breakpoints in ecosystems with a multiplicity of stable states. Nature 269:471–77.

May, R. M., and G. F. Oster. 1976. Bifurcations and dynamic complexity in simple ecological models. Univ. of Chicago Press, Am. Nat. 110:573–99.

Maynard-Smith, J. 1974. Models in Ecology. Cambridge Univ. Press, Cambridge, England.

Maynard-Smith, J., and M. Slatkin. 1973. The stability of predator–prey systems. Ecology 54:384–91.

Melchior, H. R. 1972. Summer herbivory by brown lemming at Barrow, Alaska. Proc. of 1972 I.B.P. Tundra Biome Symposium (ed. J. Brown and S. Bowen). pp. 136–38.

Meslow, E. C., and L. B. Keith. 1968. Demographic parameters of a snowshoe hare population. J. Wildl. Manage. 32:812–34.

———. 1971. A correlation analysis of weather versus snowshoe hare population parameters. J. Wildl. Manage. 35:1–15.

Middleton, A. D. 1934. Periodic fluctuations in British game populations. J. Anim. Ecol. 3:231–49.

Miller, G. S. 1912. Catalogue of the Mammals of Western Europe. British Museum, London.

Miller, R. S. 1954. Food habits of the wood-mouse (*Apodemus sylvaticus*) and the bank vole (*Clethrionomys glareolus*) in Wytham Wood, Berkshire. Saugetierk. Mitt. 2:109–14.

———. 1955a. A survey of the mammals of Bylot Island, Northwest Territories. Arctic 8(3).

———. 1955b. Activity rhythms in the wood mouse, *Apodemus sylvaticus* and the bank vole, *Clethrionomys glareolus*. Proc. Zool. Soc. Lond. 125:505–19.

———. 1958. A study of a wood mouse population in Wytham Wood, Berkshire. J. Mammal. 39:477–93.

Mock, S. J., and W. D. Hibler III. 1976. The 20-year oscillation in eastern North American temperature records. Nature 261:484–86.

Montroll, E. W. 1972. Some statistical aspects of the theory of interacting species.

In: Lectures on Mathematics in the Life Sciences, vol. 4. Some Mathematical Questions in Biology, vol. 3. American Mathematical Society, Providence, R.I.
Moran, P. A. P. 1949a. The statistical analysis of the sunspot and lynx cycles. J. Anim. Ecol. 18:115–16.
———. 1949b. The spectral theory of discrete stochastic processes. Biometrika 36:63–70.
———. 1950a. Some remarks on animal population dynamics. Biometrics 6:250–58.
———. 1950b. The oscillatory behavior of moving averages. Proc. Camb. Phil. Soc. 46:272–80.
———. 1952. The statistical analysis of game-bird records. J. Anim. Ecol. 21:154–58.
———. 1953a. The statistical analysis of the Canadian lynx cycle. I: structure and prediction. Aust. J. Zool. 1(2):163–73.
———. 1953b. The statistical analysis of the Canadian lynx cycle. II: synchronization and meteorology. Aust. J. Zool. 1(3):291–98.
———. 1954. The logic of the mathematical theory of animal populations. J. Wildl. Manage. 18:60–66.
Morse, P. M., and G. E. Kimball. 1951. Methods of Operations Research. Technology Press at MIT, Cambridge, and Wiley, New York.
Mullen, D. A. 1968. Reproduction in brown lemmings (*Lemmus trimucronatus*) and its relevance to their cycle of abundance. Univ. Calif. Publs. Zool. 85:1–24.
Myers, J. H., and C. J. Krebs. 1971. Genetic, behavioral, and reproductive attributes of dispersing field voles *Microtus pennsylvanicus* and *Microtus ochrogaster*. Ecol. Monogr. 41:53–78.
Naumov, N. P. 1934. Periodicity in fluctuation in the abundance of the common squirrel in the U.S.S.R. In Manual: Ecology of the Squirrel. KOIZ,M.
———. 1972. The Ecology of Animals. F. K. Plous, trans. N. D. Levine, ed. Univ. of Illinois Press, Urbana. (Original 1963.)
Negus, N. C., and P. J. Berger. 1977. Experimental triggering of reproduction in a natural population of *Microtus montanus*. Science 196:1230–31.
Nellis, C. H., S. P. Wetmore, and L. B. Keith. 1972. Lynx–prey interactions in central Alberta. J. Wildl. Manage. 36:320–29.
Nelson, E. W. 1909. The rabbits of North America. N. Am. Fauna 29.
Newson, J., and D. Chitty. 1962 Haemoglobin levels, growth and survival in two *Microtus* populations. Ecology 43:733–38.
Nicholson, A. J. 1933. The balance of animal populations. J. Anim. Ecol. 2:132–78.
———. 1954. An outline of the dynamics of animal populations. Aust. J. Zool. 2:9–65.
Nicholson, A. J., and V. A. Bailey. 1935. The balance of animal populations. Proc. Zool. Soc. Lond. 3:551–98.

Nicolis, G. 1972. Fluctuations around non-equilibrium states in open nonlinear systems. J. Stat. Phys. 6:195–222.

Nisbet, R. M., and W. S. C. Gurney. 1976a. A simple mechanism for population cycles. Nature 263:319–20.

———. 1976b. Population dynamics in a periodically varying environment. J. Theor. Biol. 56:459–75.

Nisbet, R. M., W. S. C. Gurney, and M. A. Pettipher. 1977. An evaluation of linear models of population fluctuations. J. Theor. Biol. 68:143–60.

Noy-Meir, I. 1975. Stability of grazing systems: an application of predator–prey graphs. J. Ecol. 63:459–81.

———. 1978. Stability in simple grazing models: effects of explicit functions. J. Theor. Biol. 71:347–80.

Ognev, S. I. 1962. Mammals of Eastern Europe and Northern Asia, vol. II. Carnivora. Trans. from the Russian. National Science Foundation, Washington, D.C.

———. 1963. Mammals of the USSR and Adjacent Countries, vol. VI. Rodents. Trans. from the Russian. Russ. Israel Program for Scientific Translations, Jerusalem.

Oldenburg, H. 1666. Some observations of swarms of strange insects and the mischiefs done by them. Phil. Trans. Lond. B Biol. Sci. 1:137.

Osgood, W. H., Preble, E. H., and Parker, G. H. 1915. The fur seals and other life of the Pribilof Islands, Alaska, in 1914. Senate Documents, vol. 6, no. 980, Washington, D.C.

Oster, G., and Y. Takahashi. 1974. Models for age-specific interactions in a periodic environment. Ecol. Monogr. 44:483–501.

Palmgren, P. 1949. Some remarks on the short-term fluctuations in the number of northern birds and mammals. Oikos 1(1):114–21.

Patterson, G. 1886. The plague of mice in Nova Scotia and Prince Edward Island. Can. Rec. Sci. 2:472–80.

Pearson, O.P. 1971. Additional measurements of the impact of carnivores on California voles (*Microtus californicus*). J. Mammal. 52:41–49.

Pease, J. L., R. H. Vowles, and L. B. Keith, 1979. Interaction of snowshoe hares and woody vegetation. J. Wildl. Manage. 43:43–60.

Pennant, T. 1784. Arctic Zoology. Henry Hughs, London.

Pennsylvania Game Commission. 1965. Pennsylvania Hunting Facts.

Peppard, L. E. 1975. Computer simulation models: applications to the study of ecological systems. Int. J. Syst. Sci. 6:983–99.

Petersen, R. L. 1966. The Mammals of Eastern Canada. Oxford Univ. Press, Toronto.

Pianka, E. 1970. On *r*- and *K*-selection. Am. Nat. 104:592–97.

Pielou, E. C. 1969. An Introduction to Mathematical Ecology. Wiley–Interscience, New York.

Pieper, R. D. 1964. Production and chemical composition of arctic tundra vegetation and their relation to the lemming cycle. Ph.D. dissertation, Univ. of California, Berkeley.

Pitelka, F. A. 1957. Some aspects of population structure in the short-term cycle of the brown lemming in northern Alaska. Cold Spring Harbor Symp. Quant. Biol. 22:237–51.

———. 1959. Population studies of lemmings and lemming predators in northern Alaska. XVth Int. Congr. Zool. Sect. X:757–59.

———. 1964. The nutrient recovery hypothesis for Arctic microtine cycles. I. Introduction. In: A. J. Crisp, ed. Grazing in Terrestrial and Marine Environments, pp. 55–56. British Ecological Society, Blackwell Scientific Publications, Oxford.

———. 1969. Ecological studies on the Alaskan arctic slope. Arctic 22:333–40.

———. 1972. Cycle pattern in lemming populations near Barrow, Alaska. Proc. 1972 Tundra Biome Symp. U.S. Tundra Biome Ecosyst. Anal. Stud.

———. 1973. Cyclic patterns in lemming populations near Barrow, Alaska. In: M. E. Britton, ed. Alaska Arctic Tundra. pp. 199–215. Tech. Pap. 25, Arct. Inst. N. Am. Wash.

Pitelka, F. A., P. Q. Tomich, and G. W. Treichel, 1955a. Breeding behavior of jaegers and owls near Barrow, Alaska. Condor 57:3–18.

———. 1955b. Ecological relations of jaegers and owls as lemming predators near Barrow, Alaska. Ecol. Monogr. 25:85–117.

Platt, T., and K. Denman. 1975. Spectral theory in ecology. Annu. Rev. Ecol. Syst. 6:189–210.

Poland, H. 1892. Fur-bearing Animals in Nature and in Commerce. Gurney and Jackson, London.

Pontoppidan, Erich. 1751, trans. 1755. The Natural History of Norway. Trans. from the Danish. A. Linde, London.

Preble, E. A. 1908. A biological investigation of the Athabaska–Mackenzie region. N. Am. Fauna 27.

Prigognine, I., and G. Nicolis. 1973. Fluctuations and the mechanism of stability. In: Proc. 3rd Int. Conf. Theor. Phys. Biol., Versailles, 1971. Karger, Basel.

Rae, J. 1890. Notes on some of the birds and mammals of the Hudson's Bay Territory. . . . J. Linn. Soc. Lond. Zool. 20:136–45.

Rescigno, A. 1968. The struggle for life: II. Three competitors. Bull. Math. Biophys. 30:291–98.

Rescigno, A., and I. W. Richardson. 1967. The struggle for life: I. Two species. Bull. Math. Biophys. 27:85–89.

———. 1973. The deterministic theory of population dynamics. In: R. Rosen, ed. Supercellular Systems. Foundations of Mathematical Biology, vol. 3. Academic Press, New York.

———. 1976. The struggle for life. I. Two species. Bull. Math. Biophys. 29:377–88.

Richardson, J. 1829. The Quadruped. Pt. I. Fauna Boreali Americana. J. Murray, London.

Richardson, L. F. 1960. Arms and Insecurity. Quadrangle Books, Chicago.

Roberts, F. S. 1976. Discrete Mathematical Models with Applications to Social,

Biological and Environmental Problems. Prentice-Hall, Englewood Cliffs, N.J.

Robinson, J. H. 1972. Astronomy Data Book. Newton Abbot, D. C. Holdings, Devon, England.

Robson, J. 1752. An Account of a Six-Year Residence in Hudson's Bay, 1733 to 1736, 1744–1747. London.

Rodin, L. E., and N. O. Bazilevich. 1967. Production and Mineral Cycling in Terrestrial Vegetation. Scripta Technica, trans. Oliver & Boyd, Edinburgh.

Rosen, R. 1968. Some comments on the physico-chemical description of biological activity. J. Theor. Biol. 8:380–86.

———. 1970. Stability Theory and Its Applications. Dynamical Systems Theory in Biology, vol. I. Wiley–Interscience, New York.

———, ed. 1973. Supercellular Systems. Foundations of Mathematical Biology, vol. 3. Academic Press, New York.

Rosenzweig, M. L. 1969. Why the prey curve has a hump. Am. Nat. 103:81–87.

———. 1971. Paradox of enrichment: destabilization of exploitation ecosystems in ecological time. Science 171:385–87.

Rosenzweig, M. L., and R. H. MacArthur. 1963. Graphical representation and stability conditions of predator–prey interactions. Am. Nat. 97:209–23.

Ross, B. R. 1861. A popular treatise on the fur-bearing animals of the Mackenzie River district. Can. Nat. Geol. 6:5–35.

Ross, R. 1908. Reports on the Prevention of Malaria in Mauritius. Waterloo, London.

———. 1911. The Prevention of Malaria. 2d ed., appendix. London.

Rowan, W. 1950. Canada's premier problem of animal conservation: a question of cycles. In: New Biology, vol. 9, pp. 38–57. Penguin Books, Baltimore.

———. 1954. Reflections on the biology of animal cycles. J. Wildl. Manage. 18:52–60.

Rudberg, S. 1968. Geology and morphology. Sømme, 1968:31–47.

Rue, L. L. 1969. The World of the Red Fox. Lippincott, Philadelphia.

Rumney, G. R. 1968. Climatology and the World's Climates. Macmillan, New York.

Russell, F. 1898. Explorations in the far north. Univ. of Iowa, Iowa City. 290 pp.

Rykiel, E. J., and N. T. Kuenzel. 1971. Analog computer model of "The Wolves of Isle Royale." In: B. C. Patten, ed. Systems Analysis and Simulation in Ecology, pp. 513–41. Academic Press, New York.

Schaffer, W. M., and R. H. Tamarin. 1973. Changing reproductive rates and population cycles in lemmings and voles. Evolution 27:111–24.

Schlesinger, W. H. 1976. Toxic foods and vole cycles: additional data. Am. Nat. 110:315–17.

Schultz, A. M. 1964. The nutrient recovery hypothesis for Arctic microtine cycles. II. Ecosystem variables in relation to Arctic microtine cycles. In: A. J. Crisp, ed. Grazing in Terrestrial and Marine Environments, pp. 57–68. British Ecological Society, Blackwell Scientific Publications, Oxford.

———. 1969. A study of an ecosystem: the arctic tundra. In: G. M. Van Dyne, ed. The Ecosystem Concept in Natural Resource Management, pp. 77–93. Academic Press, New York.

Schwerdtfeger, F. 1968. Okologie der Tiere II: Demokologie. Verlag Paul Parey, Hamburg.

Scudo, F. M. 1971. Vito Volterra and theoretical ecology. Theor. Popul. Biol. 2:1–23.

Searles, S. R. 1966. Matrix Algebra for Biological Sciences. Wiley, New York.

Selye, H. 1946. The general adaptation syndrome and the diseases of adaptation. J. Clin. Endocrinol. 6:117–230.

Seton, E. T. 1909. Life Histories of Northern Animals. Scribner's, New York.

———. 1911. The Arctic Prairies. Scribner's, New York.

———. 1929. Lives of Games Animals, vol. 1. Doubleday, Doran, New York.

Shelford, V. E. 1943. The abundance of the collared lemming (*Dicrostonyx groenlandicus* (Tr.) var. *richardsoni mer*) in the Churchill area, 1929 to 1940. Ecology 24:472–84.

———. 1945. The relation of snowy owl migration to the abundance of the collared lemming. Auk 62:592–96.

Siivonen, L. 1948. Structure of short-cyclic fluctuations in numbers of mammals and birds in the northern hemisphere. Pap. Game Res. Helsinki 1.

———. 1954a. On the short-term fluctuations in numbers of tetraonids. Finn. Found. Game Preserv. Pap. Game Res. 13:1–10.

———. 1954b. Some essential features of short-term population fluctuation. J. Wildl. Manage. 18:38–45.

———. 1956. The correlation between fluctuations of partridge and European rare populations and the climatic conditions of winter in southwestern Finland during the last thirty years. Pap. Game Res. Helsinki 17.

———. 1957. The problem of the short-term fluctuations in numbers of tetraonids in Europe. Finn. Found. Game Preserv. Pap. Game Res. 19:1–44.

Siivonen, L., and J. Koskimies. 1955. Population fluctuations and the lunar cycle. Pap. Game Res. Helsinki 14:1–22.

Slobodkin, L. B. 1954. Cycles in animal populations. Am. Sci. 42:658–60, 666.

———. 1961. Growth and Regulation of Animal Populations. Holt, Rinehart and Winston, New York.

Smith, F. E. 1952. Experimental methods in population dynamics: a critique. Ecology 33:441–50.

Smith, M. H., C. T. Garten, Jr., and P. R. Ramsey. 1975. Genic heterozygosity and population dynamics in small mammals. In: C. L. Markert, ed., Isozymes IV. Genetics and Evolution. Academic Press, New York.

Smylie, D. E., G. K. D. Clarke, and T. J. Ulrych. 1973. Analysis of irregularities in the earth's rotation. In: B. Alder, S. Fernbach, and M. Rotenberg, eds. Methods in Computational Physics, pp. 391–430. Academic Press, New York.

Smythe, C. M., and J. A. Eddy. 1977. Planetary tides during the Maunder Sunspot Minimum. Nature 266:434–35.

Solomon, M. E. 1949. Natural control of animal numbers. J. Anim. Ecol. 18:1–35.

———. 1964. Analysis of processes involved in the natural control of insects. Adv. Ecol. Res. 2:1–58.

Sømme, A., ed. 1968. A Geography of Norden. Heinemann, London.

Soper, J. D. 1921. Notes on the snowshoe rabbit. J. Mammal. 2:101–08.

Speller, S. W. 1972. Biology of *Dicrostonyx groenlandicus* on Truelove Lowland, Devon Island, N.W.T. In: L. P. Bliss, ed., Devon Island I.B.P. Project, High Arctic Ecosystem, pp. 257–71. Univ. of Alberta, Edmonton.

Strong, W. D. 1930. Notes on mammals of the Labrador interior. J. Mammal. 11:1–10.

Svardson, G. 1957. The "invasion" type of bird migration. Br. Birds 50: 314–43.

Tamarin, R. H. 1977. Demography of the beach vole (*Microtus breweri*) and the meadow vole (*Microtous pennsylvanicus*) in southeastern Massachusetts. Ecology 58:1310–21.

———. 1978. Dispersal, population regulation and *K*-selection in field mice. Am. Nat. 112:545–55.

Tamarin, R. H., and C. J. Krebs. 1969. *Microtus* population biology II. Genetic changes at the transferrin locus in fluctuating populations of two vole species. Evolution 23:183–211.

Tanner, J. T. 1975. The stability and the intrinsic growth rates of prey and predator populations. Ecology 56:855–67.

Tansky, M. 1978. Switching effect in prey–predator system. J. Theor. Biol. 70:263–71.

Tast, J., and O. Kalela. 1971. Comparison between rodent cycles and plant production in Finnish Lapland. Ann. Acad. Sci. Fenn. Ser. A, IV. Biol. 186:1–14.

Thompson, David. 1812. V. Hopwood, ed. 1971. Travels in Western North America, 1784–1812. Macmillan, Toronto.

Thompson, D. Q. 1955a. The role of food cover in population fluctuations of the brown lemming at Point Barrow, Alaska. Tran. N. Am. Wildl. Conf. 20:166–76.

———. 1955b. The 1953 lemming emigration at Point Barrow, Alaska. Arctic 8:37–45.

Thrall, R. M. 1978. Book review of An Introduction to Mathematical Models in the Social and Life Sciences. Am. Sci. 66:764.

To, L. P., and R. H. Tamarin. 1977. The relation of population density and adrenal gland weight in cycling and noncycling vole (*Microtus*). Ecology 58:928–34.

Tukey, J. W. 1965. Use of numerical spectrum analysis in geophysics. In: Statistics in the physical sciences. Bull. Int. Stat. Inst. 17(Sept.).

Tuljapurkar, S. D. 1976. Stability of Lotka–Volterra systems. Nature 264:381.

Turner, B. N., and S. L. Iverson. 1973. The annual cycle of aggression in male *Microtus pennsylvanicus*, and its relation to population parameters. Ecology 54:967–81.

Utida, S. 1955. Fluctuations in the interacting populations of host and parasite in relation to the biotic potential of the host. Ecology 36:202–06.

Verhulst, M. 1845. Recherches mathematiques sur la loi d'accroissement de la population. Mem. Acad. Roy. Belg. 18:1–38.

Volterra, V. 1926. Fluctuations in abundance of a species considered mathematically. Nature 118:558–60.

———. 1927. Variazioni e fluttuazioni del numero d'individui in specie animali conviventi. R. Comit. Talass. Italiano, Memoria 131, Venezia.

———. 1931. Leçons sur la théorie mathématique de la lutte pour la vie. Gauthier-Villars, Paris.

———. 1937. Principes de biologie mathématique. Acta Biotheor. 3:1–36.

Waldmeier, M. 1961. The Sunspot Activity in the Years 1610–1960. Schulthess, Zurich.

Walker, J., A. van Nypelseer, and W. E. Langlois. 1976. Numerical integration of a stochastic model for the Volterra–Lotka reaction. Bull. Math. Biol. 38:535–46.

Wallen, C. C., ed. 1970. Climates of Northern and Western Europe. World Survey of Climatology, vol. 5. Elsevier, Amsterdam.

Wangersky, P. J. 1978. Lotka–Volterra population models. Ann. Rev. Ecol. Syst. 9:189–218.

Wangersky, P. J., and W. J. Cunningham. 1956. On time lags in equations of growth. Proc. Natl. Acad. Sci. U.S.A. PNAS 42:699–702.

———. 1957a. Time lag in population models. Cold Spring Harbor Symp. Quant. Biol. 22:329–38.

———. 1957b. Time lag in prey–predator population models. Ecology 38:136–39.

Watt, K. E. F. 1968. Ecology and Resource Management. McGraw-Hill, New York.

———. 1969. A comparative study on the meaning of stability in five biological systems: insects and fur-bearer populations, influenza, Thai hemorrhagic fever, and plague. Brookhaven Symp. Biol. 22:142–50.

———. 1973. Principles of Environmental Science. McGraw-Hill, New York.

Weinstein, M. S. 1977. Hares, lynx, and trappers. Am. Nat. 111:806–08.

Whitney, C. 1896. On Snowshoes to the Barren Grounds. Harper, New York.

Whittaker, R. H. 1975. Communities and Ecosystems, 2d ed. Macmillan, New York.

Wielgolaski, F. E., ed. 1975. Plants and Microorganisms. Fennoscandian Tundra Ecosystems, part I. Springer-Verlag, New York.

Wildhagen, A. 1952. Om vekslingene i bestanden av sinagnere i Norge 1871–1949, J. Steenberg, Drammen.

———. 1953. On the Reproduction of Voles and Lemmings in Norway. Arbeidernes Aktietrykkeri, Oslo.

Wiley, F. A., ed. 1954. Ernest Thompson Seton's America. Devin-Adair, New York.

Williamson, M. 1972. The Analysis of Biological Populations. Edward Arnold, London.

———. 1975. The biological interpretation of time series analysis. Bull. Inst. Math. Appl. 11:67–69.

Windberg, L. A., and L. B. Keith. 1976a. Snowshoe hare population response to artificial high densities. J. Mammal. 57:523–53.

——— 1976b. Experimental analyses of dispersal in showshoe hare populations. Can. J. Zool. 54:2061–81.

———. 1978. Snowshoe hare populations in woodlot habitat. Can. J. Zool. 56:1071–80.

Wolff, J. O. 1977a. Habitat utilization of snowshoe hares (*Lepus americanus*) in interior Alaska. Ph.D. dissertation, Univ. of California, Berkeley.

———. 1977b. The role of habitat patchiness in the population dynamics of Alaskan snowshoe hares. Author's Manuscript.

———. 1978a. Burning and browsing effects on willow growth in interior Alaska. J. Wildl. Manage. 42:135–40.

———. 1978b. Food habits of snowshoe hares in interior Alaska. J. Wildl. Manage. 42:148–53.

Wood, C. A., and R. R. Lovett. 1974. Rainfall, drought, and the solar cycle. Nature 251:594–96.

von Wrangell, F. 1844. Narrative of an expedition to the Polar Sea in the years 1820, 1821, 1822, and 1823. E. Sabine, trans. London.

Wright, S. 1921. Correlation and causation. J. Agric. Res. 20:557–85.

———. 1943. Isolation by distance. Genetics 28:114–38.

———. 1960. The treatment of reciprocal interaction with or without lag in path analysis. Biometrics 16:423–45.

———. 1978. Variability within and among Natural Populations. Evolution and the Genetics of Populations, vol. 4. Univ. of Chicago Press, Chicago. 580 pp.

Wynne-Edwards, V. C. 1962. Animal Dispersion in Relation to Social Behavior. Oliver & Boyd, Edinburgh.

Yablokov, A. V. 1974. Variability of Mammals. (Nauka Publishers, Moscow, 1966.) Amerind Publ. Co., New Delhi.

Ziegler, J. 1532. Quae intus continentur. Strasbourg.

van Zyll de Jong, C. G. 1975. Differentiation of the Canada lynx. *Felis* (*Lynx*) *canadensis subsolana*, in Newfoundland. Can. J. Zool. 53:699–705.

Index of Species

General Index

(Latin names within this index refer to the Index of Species)